Moon

Moon

A BRIEF HISTORY

Bernd Brunner

Yale UNIVERSITY PRESS

new haven and london

Published with assistance from the foundation established in memory of Philip Hamilton McMillan of the Class of 1894, Yale College.

Yale University Press books may be purchased in quantity for educational, business, or promotional use. For information, please e-mail sales.press@yale.edu (U.S. office) or sales@yaleup.co.uk (U.K. office).

Designed by Sonia L. Shannon
Set in Stempel Schneidler type by Tseng Information Systems, Inc.
Printed in the United States of America.

Library of Congress Cataloging-in-Publication Data
Brunner, Bernd, 1964–
Moon : a brief history / Bernd Brunner.
    p.   cm.
Includes bibliographical references and index.
ISBN 978-0-300-15212-8 (cloth : alk. paper) 1. Moon. I. Title.
QB581.B78 2010
523.3—dc22

                                    2010015126

A catalogue record for this book is available from the British Library.

This paper meets the requirements of ANSI/NISO Z39.48-1992 (Permanence of Paper).

10 9 8 7 6 5 4 3 2 1

Contents

Introduction

The light of the sun is our essential source of energy. Without it, not only would Earth's temperature drop to an unimaginable level, but a thick crust of ice would soon cover the planet's surface. Few microorganisms, if any, would survive. If the sun disappeared altogether, our world would even lose its gravitational anchor. The world is simply inconceivable without the sun. But what would the planet be like without the moon? We might assume that the absence of our satellite would affect us less dramatically than the loss of the sun. But the more we recognize how intimately life on Earth is connected with the moon, the more unsettling the idea of life without it becomes. By stabilizing our planet's inclination, the moon has prevented it from going off on seasonal excursions that might have led to completely other consequences for the evolution of life on Earth.

Without our moon, Earth would be a vastly different place. Just *how* different it is difficult to know, but if we imagine Earth with a much weaker ebb and flow of the tides, we get an inkling of the importance of the moon's role. Life itself may not have been possible without the moon. The strong ocean tides have created pools considered essential for complex biological systems to arise. According to the theory now generally accepted, a body with the dimensions of Mars collided with our planet, flinging off the disk of material that created the moon and gave Earth its axial tilt. What would our planet be like if such an incident had never taken place?

Although there are limits to the credibility of speculative history, some scientists find such thought experiments irresistible. The American astronomer and physics professor Neil F. Comins, for example, draws an elaborate comparison between an Earth on which no impact has occurred, which he calls Solon, and Earth as we know it. According to Comins's hypotheses, Solon rotated much faster—possibly at three times Earth's current speed—which would cause correspondingly stronger movements in the planet's atmosphere. Tall trees or delicate, large leaves would be a rarity in such a world, as would fragile animals with long legs or wings. Humanoid creatures could exist, but their appearance would be quite different. If Comins is right, the moon has therefore played a pivotal role in our development as a species. Although the precise extent of its influence can probably never be determined, Earth's lifeless satellite plays a central part in who we are—along with the sun, the atmosphere, our oceans, and the animals and plants living beside us. The moon's history is intimately related to the Earth's.

Yet the moon's central role in life on Earth is difficult to square with the simple facts we know about it. Our moon is a bleak, gloomy, lifeless celestial body just a quarter of Earth's size and one-eighty-first of its weight, with a gravitational pull of only one-sixth. It rotates on its own axis once approximately every twenty-eight days, which is very slow compared with the current twenty-four-hour rotation period of the Earth. Its surface is a bit larger than Africa's and Australia's combined. Its very thin atmosphere means that there is no sound and, without a means to retain the warmth of the sun, the surface temperature fluctuates considerably.

In some ways, viewing and scrutinizing the moon gives us a look back to the beginning of the solar system. The position of the moon relative to Earth has changed, though. According to computer models, two billion years ago the moon would have been 24,000 miles away from Earth, orbiting it 3.7 times per day, and causing tides up to a thousand times higher than those observed today. Now at an average distance of 238,900 miles from the Earth—or the equivalent of thirty times the Earth's diameter—the moon is losing energy, slowing down, and receding from us as its orbit expands by 1.5 inches each year, or 250 feet in two thousand years. The tides act as a brake on the rotational speed of both the moon and the Earth. At some distant point in the future, the sun will engulf the Earth and the moon. But for now, the sun will continue to shine, the wind to blow, the seas to surge.

These physical facts and figures alone do not explain what the moon means to us. The impor-

fig. 52. Wie der Mond um die Erde kreist und ihr immer dieselbe Seite zukehrt.

The moon circles the Earth, while always exposing the same side. The tension of the cord stands for the attraction of the Earth, locking the spin of the moon.

tance of the moon has less to do with its proximity to Earth than with its centrality in the human imagination. The moon's face has been known to inspire admiration, sadness, joy, even longing, and, under certain conditions, fear. Even though we may think we can lasso the moon, it often eludes our grasp. Close, yet really far away— the moon is a paradox. And when we study it, we are also studying an aspect of ourselves.

If Earth had been shrouded in clouds, objects in the sky would never have evolved into symbols, but because we can see the moon wax and wane each month, for example, we take meaning from its periodic journey between complete darkness and a bright, full disk. The moon's nearness also encourages us to ponder what might be "out there." Are there other worlds among the distant stars? Is there another sphere similar or not so similar to ours? The moon's proximity also made it natural that we should make it the destination of our first venture beyond our home planet.

The moon is the most examined object in the sky. For millennia it has been an enigma, a *luna incognita*. The Greek writer Aeschylus (525–456 B.C.) saw in it "the eye of the night." Lucian of Samosata (ca. A.D. 120–185) wrote, "I found the stars dotted quite casually about the sky, and I wanted to know what the Sun was. Especially the phenomena of the Moon struck me as extraordinary, and quite passed my comprehension; there must be some mystery to account for those many phases, I conjectured." For a long time, the moon had counted among the seven planets moving about a fixed Earth, along with Mercury, Venus, the sun, Mars, Jupiter, and Saturn. In the seventeenth century, as the heliocentric idea of the universe gain wider acceptance, the moon was relegated to a less important position. Now it was just a satellite, not even unique, since most of the planets circling the sun have one or more moons.

This book is a brief history of the imprint the moon has left on the human imagination and of the enduring fascination it has provoked. It is an appreciation of the contributions of various cultures—and of both popular and scientific traditions—in shaping our sense of the moon. An appreciation, too, of the moon's unique power to inspire our capabilities for invention and to nourish our drive to self-understanding. Over the millennia the moon has been the focus of a vast array of human practices. This book is devoted to many questions regarding the moon's role in our lives and the lives of those who went before us. How, for example, has the moon been used to structure time? What kinds of life have both scientists and writers imagined on the moon? How has the moon's origin been accounted for? Why do some people—all the evidence notwithstanding—still claim that the moon landings never happened? My hope is that these investigations, taken together, will be more than a motley collection, that they will provide a sense of the amazing continuity, from the first imagined lunar journeys to the Apollo program, of the moon's eternal place in the life of humankind.

Moon

Gazing at the Moon

While the sun is too bright for us to look at it directly, the moon lends itself to gazing and contemplating. Over the course of about one month, the moon makes a perceptible trip through the sky. Its phases are more easily distinguishable than its motion. On the third day after the new moon, its visible surface starts to take the form of a thin semicircle, easily lending to a comparison with a pair of horns or a boomerang. The next night it will be higher above the western horizon than the night before, and not as thin. It also sets later as its month moves on. It develops into a half-moon. During the next seven to eight days, its light increases until its image becomes circular. At the time of the full moon, it is directly opposite the sun, so that the latter illuminates the full surface of the moon visible from Earth throughout the whole night. Next the moon goes again through the same shapes as before: now from oval to last quarter, the phases of the waning moon mirror those of the waxing moon.

The last quarter diminishes, further taking the shape of a crescent, the horns of which are raised on the side farthest from the sun. At some point, by the twenty-seventh day, the moon is visible only for a short period of time before sunrise. During the last hours of darkness it can still be

Moon gazing was all the rage in nineteenth-century Europe.

seen, but it's clearly fading. It approaches the sun and finally gets lost in its rays. The moon is actually part of the daytime sky as often as the nighttime sky, even if it is hardly noticed or sometimes mistaken for a soft cloud. Finally, for the duration of three days, the moon is no longer visible in the sky, neither at day nor at night—except during a solar eclipse, when a sliver of the moon is visible. These regularly recurring phases are a consequence of the moon's movement around the Earth, and the phase of the moon as seen from the Earth is always complementary to that of the Earth from the moon. The moon moves along its path about thirteen times faster than the sun, covering the distance in four weeks that the sun travels in a year.

During the full moon, details of the lunar surface are indistinguishable; even the mountains of the moon barely cast a shadow. Tycho, for example, the brightest and most conspicuous crater on the moon—named after the Danish astronomer Tycho Brahe (1546–1601)—is eighty-five kilometers wide and, at perhaps 108 million years old, the youngest of the craters on the near side of the moon. It was formed a long time before humans walked the Earth, but dinosaurs could have observed the moment of impact. As the moon waxes,

The development of the lunar phases as illustrated in a print by John of Sacrobosco (also known as John of Holywood) from *Tractatus de sphaera,* an important astronomy book of the Middle Ages

APPENDIX.

the appearance of Tycho experiences a remarkable transformation. Initially it is a great gaping crater—reminiscent of the Greek root of the word, meaning "cup" or "bowl"—then it slowly grows to become the epicenter of a complex system of rays extending hundreds or thousands of kilometers into the area around it. It is said that Tycho makes the moon resemble a peeled orange. The lunar scientist Thomas Gwyn Elger (1836–1897) called it "the Metropolitan crater of the Moon."

Long before the discovery of the telescope, humans pondered the particular pattern of lighter and darker areas on the lunar surface, much as they have always studied the shapes of clouds in the sky. The human face imagined on the lunar surface was probably the most ancient, and certainly the most anthropomorphic, perception. In this case, the major dark spots on the lunar surface—the maria—would be associated with certain facial features like eyes, eyebrows, nose, cheek, and lips. When we look at the moon with the naked eye, the maria vaguely suggest a human face. Sometimes the pattern of dark and bright spots is identified with the contours of a woman's face, with her hair bound up on the top of the head. Armed with a little imagination, observers could project a vast repertoire of images on the pattern of dark and light areas: from a broad grinning face or a rabbit with long ears to a crab or even a man with a dog. Often they would discern the features of the proverbial Man in the Moon.

The Tycho crater and the surrounding area

Among the most obvious phenomena related to our neighbor in space are moonrise and moonset. Both can be impressive, though the amount of light involved cannot produce a colorful optical effect as spectacular as the interplay of intense red and orange hues at sunrise or sunset. The moon appears much larger—double or even triple in size—when it rises or sets, dwarfing the houses and trees surrounding it. After its ascension to the sky, this impression disappears. Sev-

A hare in the moon?

eral possible explanations have been advanced to account for the "moon illusion." One is that the close juxtaposition of objects such as houses or trees with the brightly illuminated moon fools us into exaggerating its size compared with the foreground objects. But the illusion may also plausibly be related to the perceived distance of the moon: when it is on the horizon, the brain interprets it to be farther away than when it is above us.

As incredible as it may seem, the French astronomer Frédéric Petit, then the director at the Toulouse Observatory, was convinced he had discovered a second moon with an elliptical orbit while looking through his telescope early one evening in 1846. Jules Verne took up this idea in his book *From the Earth to the Moon*. We cannot reconstruct how Petit arrived at such a conclusion, but a small line of astrologers after him claimed to have seen second or even multiple moons. One was the German Georg Waltemath, who, shortly before the close of the nineteenth century, reported seeing an entire group of midget moons. Two decades later, Walter Gornold—another German—gave the name Lilith to what he claimed was a dark moon visible only when it crossed the sun.

How about other phenomena associated with the moon? Both myth and history bear witness to the overwhelming sensations inspired by the beauty of the full moon. It provided the backdrop for holy marriages between gods and goddesses, as well as for coronations and dancing rituals.

Gautama Buddha is believed to have attained enlightenment on the day of the full moon while sitting under the Bodhi Tree. Many took the full moon as an occasion for unusual behavior. The Chuckchee shamans from northeast Siberia reportedly undressed and exposed themselves to its light, thereby obtaining magical powers. It is doubtful whether we should credit Martin P. Nilsson's assertion that "half Africa dances in the light in the nights of full moon." On the other hand, we know that the boys and girls of the Shona, an ethnolinguistic group centered in Zimbabwe, still like to dance by the light of the full moon to the sound of drums and rattles.

In contrast, the absence of the moon from the sky was frequently accompanied by the fear—contrary to empirical evidence—that it would die, never to return. The Aztecs of central Mexico, for instance, believed that they recognized death in the dark of the moon. Sometimes the phase of the new moon was perceived as a liminal period, a time of prayer for the moon to return. And when the silvery disk finally appeared in the sky, it was greeted joyfully, with exclamations of salvation.

Other spectacular lunar phenomena include eclipses that are caused by or that affect the moon. The word *eclipse* goes back to the Greek *ekleipsis,* meaning "omission" or "abandonment." A total solar eclipse counts among the grandest and most dramatic sights in nature. Solar eclipses occur when the moon happens to pass between the Earth and the sun, with the moon fitting more or less over the sun's face. The phenome-

In this illustration by the French satirical illustrator Grandville (1803–1847) the solar eclipse is represented as a conjugal embrace of the sun and moon (1844).

non, which can last for up to seven minutes, is possible because the sun is about four hundred times larger than the moon and four hundred times as far away—a most curious coincidence. But this kind of eclipse is visible only within that narrow part of the Earth's surface located in the moon's shadow. In contrast, a lunar eclipse, which occurs when the Earth passes between the sun and the moon, lasts for several hours and can be seen from any point on Earth where the moon is above the horizon at the time.

The abrupt turn to blood-red, or "dying," of a full moon must have alarmed early humans, often provoking fear even when it could be predicted. An eclipse has often been interpreted as a suspension of the natural order. The Maasai people of east Africa are reported to have thrown sand into the air during an eclipse. Some North American Indians are said to have banged and rattled pots and pans (and probably drums before pans became available) or shot flaming arrows in the direction of the moon to kill the predator consuming its light. People from the Orinoco region of Venezuela buried their fire under the ground in case the moon's fire went out.

During the minutes before a total solar eclipse, with sunlight arriving only from the edge of the sun, colors turn more intense and shadows more distinct. As the temperature drops, an eclipse wind is produced, and shortly before the total eclipse so-called shadow bands appear: shimmering dark lines produced by atmospheric temperature cells produced by the remaining rays of the sun. During a solar eclipse all objects on Earth assume an unforgettable pallor, often described as mainly olive green with tinges of copper.

The French Catalan astronomer François Arago (1786–1853) described the emotions of the witnesses to a total eclipse of the sun that he observed in the eastern Pyrenees on July 8, 1842. Nearly twenty thousand people had assembled with smoked glasses, only the sick staying inside, and

> when the Sun, reduced to a narrow thread, commenced to throw on our horizon a much-enfeebled light, a sort of uneasiness

LOOKING AT THE ECLIPSE, October 19, 1865.

BURNING OF THE MICHIGAN CENTRAL FREIGHT DEPOT AT DETROIT, MICHIGAN, October 18, 1865.
[Sketched by L. T. Ives.]

SHIPWRECKED PASSENGERS AND CREW OF THE BRIG "TITANIA."—Sketched by Captain Frazer.—[See Page 695.]

A solar eclipse that passed over the North American continent on October 19, 1865, shared newspaper space with a fire and a shipwreck.

took possession of everyone. Each felt the need of communicating his impressions to those who surrounded him: hence a murmuring sound like that of a distant sea after a storm. The noise became louder as the solar crescent was reduced. The crescent at last disappeared, darkness suddenly succeeded the light, and an absolute silence marked this phase of the eclipse, so that we clearly heard the pendulum of our astronomical clock. . . . A profound calm reigned in the air; the birds sang no more. After a solemn waiting of about two minutes, transports of joy, frantic applause, salute, with the same accord, the same spontaneity, the reappearance of the first solar rays.

According to the French science writer Camille Flammarion (1842–1925), in Africa on July 18, 1860, men and women were seen either praying or fleeing to their dwellings. "We also saw animals proceeding towards the villages as at the approach of night, ducks collected into crowded groups, swallows hurling themselves against the houses, butterflies hiding, flowers—and notably those of the Hibiscus Africanus—closing their corollas," Flammarion writes. On July 28, 1851, the sky was too cloudy to yield an astronomical observation of any kind in the town of Brest in Belarus, so the astronomer Johann Heinrich von Mädler focused his attention on the fauna. Except for the horses, all animals showed some kind of restlessness, beginning fifteen minutes before the actual eclipse. Geese and ducks fell asleep, and chickens scurried to their roosts. The aurochs, one of the last of its kind, became uneasy, soon hiding in the thicket and uttering a call that had rarely been heard. Before the total solar eclipse passed over north India on October 24, 1995, where it happened to coincide with the annual Festival of Light or Divali, astrologers recommended avoiding the shadow of the moon while it crossed the sun to circumvent misfortunes. The eclipse lasted less than a minute and was visible on a path less than twenty-seven miles wide, but many observers fasted or jumped into rivers to be cleansed. A Western traveler observed the eclipse from Khanua, Rajasthan: "I noticed it had begun to darken in the West. The clear sky turned a deeper blue than normal. As more and more of the Sun disappeared, the colors changed and gave the landscape an evening glow. Somebody described it as 'spooky.' Even the villagers quietened down as a chill filled the air. The light was very eerie and it continued to darken, the light changing as I watched. Two dogs sped across the fields heading for home, their tails between their legs. Parakeets squawked and circled uncertainly. The monkeys had disappeared. I could see the black lunar shadow approaching rapidly. I looked up to see the thin crescent of the Sun condense into a point, flicker, and go out."

In some instances anticipation of an eclipse has been used to increase one's credibility and ad-

vance an agenda. When Christopher Columbus was on his fourth trip to the New World, worms in the timbers of his ship caused it to leak and become unstable. He had to land in St. Anne's Bay, Jamaica, to repair it. Ultimately, he and his crew had to spend more than a year on the island, whose indigenous people refused to defer to the Europeans. Finally, Columbus, having calculated that a total eclipse of the moon would occur on February 29, 1504, gained an advantage. The night before this event, he called a meeting of the tribal leaders, invoked the Almighty, and warned that if the natives did not cooperate, the moon would disappear from the sky. When the warning came true, the terrified natives begged Columbus

Christopher Columbus and the Jamaicans at the moment of the lunar eclipse to bring the moon back, providing his crew with food and assistance.

Thanks to precise knowledge of the motion of the sun, moon, and Earth, it is possible to calculate eclipses for thousands of years, in the past as in the future. The Austrian astronomer and mathematician Theodor von Oppolzer (1841–1886), in his *Canon of eclipses* (1887) accomplished the feat of compiling eight thousand solar and fifty-two hundred lunar eclipses covering the time span from 1200 B.C. until A.D. 2161.

Eclipses are rare, but the very notion of rareness evokes another lunar phenomenon. To say that something happens "once in a blue moon" is to say, literally, that it happens as often as a second full moon sneaks in during a calendar month. In an unrelated phenomenon, the moon may in fact appear bluish when the Earth's atmosphere contains a high concentration of very small smoke or dust particles. Particles measuring about one micron in diameter—thus wider than the wavelength of red light—strongly scatter red light, while allowing other colors to pass, so that white moonbeams shining through the clouds emerge as blue and sometimes as green. This phenomenon has also been described as resembling electric glimmer. When the volcano Krakatoa erupted in 1883, plumes of ash rose to the top of Earth's atmosphere; as a result, the moon appeared to be blue for about two years. Forest fires can have the same effect.

Among the rarer optical effects associated with the moon are also moonbows or lunar rainbows.

Although the cause is the same as for a sun's rainbow—light refracted through water droplets—the moon's bow is much weaker, usually appearing as less brightly colored and less discernible in the sky. After crossing from Nassau to Miami on June 16, 1938, the geographer Armin Kohl Lobeck of Columbia University delivered a dramatic account in *Time* magazine. "Tumultuous trade wind clouds towered to gigantic heights and there were occasional squalls of rain. About 11 o'clock, when the moon was well up in the southeast sky, the rainbow appeared in the northwest, where a thunderstorm was in progress. The prismatic colors were fairly distinguishable. The arc was complete, the two ends dipping into the sea." There are also so-called moon coronas produced by high, thin clouds, forming a close fringe around the moon. A lunar halo, a large colored ring surrounding the moon, is sometimes visible when there are ice crystals in the upper atmosphere. Moon pillars, pale shafts of light that extend out either above or below the moon, can be seen when the moon is rising or setting near the horizon. They appear when ice crystals reflect light forward from a strong light source such as the moon.

A number of people claim to have seen red glows, flashes, glows, mists, obscurations, temporary colorations of the lunar soil, and shadow effects when looking at the moon. Claims of such sightings, typically discernible only during short intervals, go far back in time, with some having been observed independently by several

A moonbow

witnesses or reputable scientists. In fact, lunar transient phenomena or LTPs, as they are now called, continue to be seen. Some topographical formations, such as the surface of the lunar impact crater Aristarchus and its satellite craters, seem to be particularly prone to the phenomenon, accounting for a third of all such observations. Lunar missions have established a higher occurrence of alpha particles caused by the emission of radon-222 from Aristarchus. This may be the source of the transient light phenomena observed.

LTPs are hard to analyze, though, in part because they are irreproducible. Reports of LTPs rarely make it into scientific publications, and many events probably arise from causes in the terrestrial atmosphere. To be considered as a genuine lunar phenomenon, an event would have to be observed from two places on Earth at the same time—a hard task for something no one can predict.

A lunar halo

If LTPs are somewhat dubious and irregular, impacts are undeniable, and they occur continually on the lunar surface. The most common impacts are those associated with micrometeorites. Impact flashes from such events have been detected simultaneously at multiple Earth locations. Most lunar scientists acknowledge that such transient events such as the emission of gases from the surface and impact cratering do occur. The controversy lies in the frequency of such events.

A category of another order are "observations" of life forms on the moon. A digression is in order. The idea of lunar life has been part of the human imagination for millennia, long before the invention of the telescope, but the English clergyman John Wilkins (1614–1672) was among the first modern scientists to purport such a view. "'Tis probable there may be inhabitants in this other World," Wilkins wrote in *The Discovery of a World in the Moone* (1638), "but of what kinde they are is uncertaine." Johann Hieronymus Schröter (1745–1818), a German astronomer known for his elaborate drawings of Mars was, "fully convinced that every celestial body may be so arranged physically by the Almighty as to be filled with living creatures." He attributed color changes he observed to cultivation of those areas, and he speculated that signs might indicate "some industrial origin—furnaces or factories of the inhabitants of the moon." The "pro-selenites" also included the eccentric German refugee Sir William (Wilhelm) Herschel (1738–1822), who, in his paper *Observations on the Mountains of the Moon,* published in 1780 in the journal of the Royal Society in London, claimed to have seen "forests" on the surface through his giant telescope and insisted that the habitability of the moon was almost certain. Furthermore, Herschel expressed what could count as one of the most daring statements about the relationship between moon and Earth: "Perhaps—and not unlikely—the Moon is the planet and the Earth the satellite! Are we not a larger moon to the Moon, than she is to us?" He also left no doubt that he would prefer the moon as his preferred habitation.

The absence of a lunar atmosphere has long cast doubt about the possibility of life on the moon. Even early observers noted that stars eclipsed by the moon don't suffer any visible change in their shape or color as they pass into or out of the line of sight to the moon. Still, the notion of life on the moon remained seductive even to scientific minds. A notable figure among the ranks of the moon gazers with particularly vivid imagination is Franz von Paula Gruithuisen (1774–1852). A Bavarian physician who first became known as a pioneer of a less invasive surgical method to remove bladder stones, Gruithuisen later became a professor of astronomy.

Although he was aware of the different temperatures and gravitational conditions on the moon, Gruithuisen claimed in a lengthy article, *On Lunar Inhabitants and Their Colossal Artistic Artifacts* (1824), to have observed manifold changes in color in the thin "lunar atmosphere"—"clouds and mists" not only diffusing on the lunar surface, but also keeping it warm and enabling floral growth. "Some plants such as cress would bear fruit under these circumstances even on the Moon," he wrote, believing that fruits ripened much faster there. He sternly postulated advanced life on the moon, meticulously hunting down traces of "understanding inhabitants." Since heat develops poorly on the lunar surface, he thought it would be impossible for the inhabitants to heat freestanding houses and imagined that they lived underground, where they could go without heating, "even though one cannot deny [their] ability to make fire or to heat in the case of neediness." Gruithuisen even claimed that he had seen a lunar city and a star-shaped "temple." The ashen light he discerned, so he assumed, might be caused by fire festivals celebrating either changes in government or religious periods. He speculated on the character of the lunar surface and said that scientists would discern many more details once they had the technical means to do so. He concluded that "dwarf-like instruments will only produce dwarf-like steps; only with the help of giant telescopes will it be possible to make giant steps, sufficient diligence assumed." Gruithuisen's "discoveries" were clearly over the top,

Franz von Paula
Gruithuisen

even for those contemporaries who had not entirely given up on the idea of life on the moon, but they helped him gain an appointment as a professor of astronomy at the University of Munich. They also caused him to develop a strange ambition to make contact with the selenites. Assuming that moon dwellers share our understanding of mathematics, he advocated building a vast geometrical structure in Siberia to attract their attention. Sadly, this project never became a reality.

Even though more and more respectable astronomers clearly declared the impossibility of life on the moon, speculation that there could be *some* sort of life didn't simply vanish overnight. The popular imagination continued to follow its own laws. Even as the twentieth century approached, Camille Flammarion clung to his belief in life on the moon. Well aware of the limitations of telescopes, he wrote: "Now, I ask, what can be distinguished and recognized at such a distance? The appearance or disappearance of the pyramids of Egypt would probably pass unnoticed." From his travels in balloons, he was aware that from a distance, a visitor might not believe in life on Earth. "If, then, the Earth seems like a dead world when seen from only a few miles' distance, what is it but illusion to assert that the Moon is truly a dead world, because we view it at 120 miles or more?" he wrote in *Popular Astronomy* (1894). Even using the highest magnification available at the time, he assumed, signs of life would not be visible. Although Flammarion knew that the

Earth and moon have very different climates, he attributed the occasional "fogs, mists, vapours, or smoke" he observed to human origin. Eventually, he began to content himself with the idea that simpler forms of life existed on the moon. "Why should we suppose that there is not, on this little globe, a vegetation more or less comparable with that which decorates ours?" he asked. "Thick forests, like those of Africa and South America, may cover vast extents of land without our being able to recognize them. On the Moon they have neither spring nor autumn, and we cannot trust to the variations of tint of our northern plants, to the verdure of May and the fall of the yellow leaves in October, to strictly typify that the lunar vegetation should show the same aspects, or should not exist. . . . Do there exist on the Moon passive beings analogous to our vegetation?"

Later on, speculations about life on the moon remained a specialty for a small caste of stubborn moon gazers. The American astronomer William Henry Pickering (1858–1938) believed the more or less sharply defined white spots scattered over the lunar surface to be ice fields, and he claimed to have seen snowstorms on a peak of the well-known pinnacle Pico and even blizzards north of the crater Conon, near the highest part of the moon's Apennines. During some phases of the lunar day he observed a greenish coloring in the center of the crater Grimaldi that he mistook for vegetation. He hypothesized that changes in the appearance of the lunar surface—movement of small dark areas—were due to "lunar insects." He

Camille Flammarion

even went so far as to define their size: similar to tropical red ants, but not exceeding the size of the locusts devastating crops in Africa. Given the difference of the environment, Pickering didn't expect them to resemble any animals found on Earth.

As late as 1960 the astronomy writer Valdemar Firsoff asked, "Is the Moon a museum piece from a geologically remote past, preserved in the vacuum of space as though in a labeled glass case? . . . Has perhaps life secured a foothold on our companion world?" With an enthusiasm bordering on obsession, he tried to revive the idea that there might be life on the moon. He regarded as likely that the moon has an atmosphere made up of water vapor, carbon dioxide, and "heavy vapours" of "volatile substances" that develop "like some sealed aquaria, within walled enclosures, clefts, and hollows, in the depressed portions of the maria, and other, less-well-defined, volcanic regions."

Firsoff reminds us that, in earlier times, Earth's atmosphere had very different constituents. Its animals would have had respiratory organs adapted to these conditions, as still do, for example, deep-sea fish that can survive in complete darkness and under high pressure. Firsoff recognized the extent of temperature fluctuations on the lunar surface, but he didn't see these factors as standing in the way of life. He pointed to the fact that on Earth, lichens and rotifers, distant relatives of the spiders, survive well below the freezing point, and that, at the other extreme, the dry sand of the Sahara Desert is full of microbes, while some protozoa survive even in boiling hot springs. Firsoff theorized that "vital parts of perennial plants may be hidden underground" and assumed that "lunar plants may not depend on the supply of gas from the atmosphere at all, or only to a small extent, obtaining all or most of their needs from the gas marsh below and reaching out beyond it only for the energy of the sunrays required for photosynthesis or similar processes."

As late as 1968 Arthur C. Clarke (1917–2008), one of the great visionaries of science fiction, speculated, "If life ever got started on the Moon — perhaps in some long-vanished lunar sea — it may still be there. Any biologist worth his salt could design a whole menagerie of plausible Selenites, granted the existence of a few common chemicals on or below the lunar surface." For skeptics Clarke invoked the example of the deserts of the American Southwest, which appear to be barren from the air but are really seething with life, as Walt Disney's documentary *The Living Desert* (1953) had impressively shown.

While some eager believers may have seized on these remarks when they were written, such fantasies finally came to a sudden end in 1969 . . . when humans set foot on the moon.

CHAPTER TWO

Moon of the Mind

Like an ongoing discussion, our understanding of the moon is constantly overlaid by new discoveries and events. Each generation has a collective perception of what the moon is, means, and symbolizes. Any effort to reconstruct this shared field of meaning from a past time reminds us that, as the historians of science Stephen Toulmin and J. Goodfield put it in *The Fabric of the Heavens* (1961), we are "confronted not by unanswered questions, but by problems as yet unformulated, by objects and happenings which had not yet been set in order, far less understood." In other words, there is not one moon but many, each particular to a different culture. So to see the moon from the perspective of earlier peoples, we must try to shed all we have learned about it—at least for now. In so doing we may better discover the place of the moon in our forebears' myths, in their stories about themselves, and in their attempts to order the flow of time.

Where to begin? We may want to start with the moon and the sun, the two brightest objects in the sky. If we follow their peregrinations over the course of the day and night, we can see one as the death of the other.

It hardly comes as a surprise that both the

In the Islamic world, the appearance of the lunar crescent was trumpeted to the people.

Caricatures of sun and moon

moon and the sun are core elements of the world's ancient religions—in fact, in many instances they were held to be the two principal sky gods. How has the relationship between these two most conspicuous celestial actors been understood? In old stories across cultures they were often characterized in human terms. They were brother and sister; a "mismatched" couple exhibiting very different personalities; or married souls forever fighting and unable to reconcile their differences. In the Rig Veda, a sacred Hindu text that dates back some four thousand years, a hymn celebrates the marriage of the Moon God with the Sun Goddess. In medieval Provence, among the Jews of Avignon, the moon was considered to be a bedraggled or evil sun—the sun of the wolves or of the hares.

In some cultures—in warmer climates, for example, like India, Mesopotamia, and Egypt, where the sun was perceived as an enemy rather than a life-giving force—the moon took precedence over the sun as an object of veneration. In temperate zones, humans seem to have realized early on that the sun's warmth was the major factor both in plant growth and in determining the seasons. Here the moon began to be associated more often with cold and darkness. Both science and religion contributed to the increasing status of the sun: astronomers revealed that moonlight is merely reflected sunlight, and the emergence of Christianity brought the belief in a transcendent God who could be symbolized by the sun.

Despite this shift, the moon remained more accessible, better suited to human measure than the sun, with its full and radiant energy. If the sun is a metaphor for unfathomable God, the moon can be interpreted as a god in a form intelligible to humans.

We have seen how the "face" of the moon was interpreted. Only a small step was required for stories to evolve from these images. A German tale tells of a man who went to collect wood in the forest, even though it was a Sunday, the day of rest. Immediately sent to the moon for his sin, he has been visible there ever since, a warning to other humans that they too could end up imprinted on the moon's face forever should they dare commit some unlawful act here on Earth.

The native people of New Zealand, the Maori, have a story that symbolizes the influence of the moon on the rain and on the waters of the Earth. In the patterns on the moon they see a woman with a bucket. This woman, Rona, was the daughter of the sea god Tangaroa, whose task it was to control the tides. One night she was carrying a bucket with stream water home to her children when the path darkened—the moon had disappeared behind the clouds. Rona, continuing her walk in the dark, tripped over a root and made some unkind remarks about the moon. From this point on Rona's people were cursed. The moon also retaliated by snatching away Rona, who continues to live in the sky. When Rona upsets her bucket, they say, there will be rain.

There is some evidence that the moon not only

An engraving
in the cave of
La Mouthe

had a firm place in the life and mind of certain peoples but may also have been understood as a living being itself, or regarded as intimately connected with certain animals. Given the absence of written documents, we must depend on the association of symbols and imagery found in certain artifacts that only suggest what significance the moon had. For example, the Venus of Laussel, which dates back to 25,000 B.C., is an engraving on limestone found in the entrance of a cave in the Dordogne area of France. In its right hand the nude female figure holds something which could be the horn of a bull but could also be a half-moon. It is carved with thirteen notches, which may refer to the number of lunar months (and menstrual cycles) in a year, especially since the figure's left hand points toward her womb. Farther east an association of the bull, pregnancy, and the moon is evident in religious shrines at Çatal Höyük, a central Anatolian settlement from the Neolithic Age, approximately 8,000 years ago. Several bulls' heads are depicted in plaster relief on the walls, along with a mother goddess, whose horns resemble the lunar crescent. In an engraving of a running bull found in the cave of La Mouthe in southwestern France, the two horns not only are larger than life but are contorted, as if to invoke more realistically the waning and waxing moon.

Sculptures of clay and bronze symbolizing bulls and the moon have been found dating back to the Bronze Age, which indicates that even then humans concerned themselves with the sky. These include the beautiful Nebra sky disk, thirty-two centimeters in diameter, which was found in Germany and dates back to ca. 1600 B.C. Among the gold ornaments on it is one that can be interpreted as a lunar crescent or an eclipsed sun. Some researchers go so far as to see the disk as evidence that the Bronze Age people already used a combination of solar and lunar calendars.

In Mesopotamian and Assyrian mythology, the god of the moon was called Sin, symbolized both by the crescent and by a strong bull with perfectly shaped limbs and distinct horns. The double axe with its concentric rings, found not only in Greece but also in the Middle East and at pre-Columbian sites, also may symbolize half-moons.

According to the poet William Butler Yeats (1865–1939), the moon "is the most changeable of symbols, and not just because it is the symbol of change." It is not difficult to see how the moon could have become a symbol for both transience and rebirth in many early cultures. References to the moon as "new," "young," and "old" reflect this pattern. Myths related to the moon are often characterized by paradoxes. As often as the moon is seen as the source of renewal, some cultures saw it as the possible source of death. For the Maori, the moon was the "man eater"; the Tatars of central Asia thought that a man-devouring giant lived in the moon; and the Tupí, at the time of colonization by the Portuguese one of the largest ethnic groups in what is now Brazil, are said to have believed that "all baleful in-fluences, thunder and floods proceed from the Moon." Many cultures share the idea that we fly to the heavens at the moment of death. In this context, the moon often becomes the first stage of such a journey. According to the Upani-shads, the ancient Hindu scriptures, for example, the deceased could take two different paths. Both passed by the moon, but one returned to Earth, while the other continued to the sun, where the union with Brahman was achieved. Arriving at the sun thus signified the end of the cycle of re-incarnation. Often, the moon was imagined as a gate to another world, a mediator between the Earth and the sun, or a transitional place to an eternal world. Some Buddhist monasteries have "moon gates," thresholds of passage to an al-together different reality.

And what of the moon's gender? The list of peoples for whom the moon is considered male is long indeed; it includes, among others, the Ainu, Anatolians, Armenians, Australian aborigines, Balts, Basques, Finns, Germans, Hindus, Japa-nese, Melanesians, Mongolians, Native Ameri-cans of the Pacific Northwest, Persians, Poles, and Scandinavians. In some Slavic folklore traditions, such as those of Russia and Bulgaria, the moon was called "father" or "grandfather." In most cases in which the moon was male, the sun was female.

The gender of the moon has not been a con-stant in all cultures. In the English language the moon was masculine up until the sixteenth

century, an idea retained with "the man in the moon." In other cultures, though, the moon has been assigned feminine attributes—though it is useful to remember that differences between "feminine" and "masculine" have not always been clear-cut and immutable. In India, for example, the word for the moon was feminine, coming from the same root as the words for mother and spirit, but the moon was brought to life as the god Chandra (both a boy's and a girl's name today, the difference only in the way it is pronounced), and sometimes Chandrasasin— the god of the moon with the hare. Today on the Indian subcontinent the moon is associated with calmness and softness, and it is common for adolescent girls to display their love for the moon. In the myths of some cultures the moon can even be male in the waxing and female in the waning.

The Esala Perahera, a procession in the honor of Buddha at Kandy in ancient Ceylon, in tune with the moon

It is male for some North American Indian tribes and female for neighboring tribes. In China, on the other hand, it is difficult if not impossible to operate with these adjectives, because sun and moon are perceived as sexless. However, they can be associated with the opposing principles yang (luminous and warm—Sun) and yin (shadowy and cold—moon).

As far as we can reconstruct beliefs today, in the Paleolithic and Neolithic periods mother goddesses were related to both the moon and the Earth. The Egyptians had various lunar myths, often connected with fertility. They venerated Isis, who gave birth to Horus, the god of the sun. Their god Thot was also connected with the moon. Sin, the Sumerian moon god and patron of the sacred city of Ur, combined both womb and bull. In ancient Greek mythology, the moon is associated with the goddess Selene, but also with Artemis and Hecate. Selene is Greek for "moon" and also the source for the word *selenology,* the encompassing term for lunar study, and for the name of the element selenium. Later, Luna and Diana were the goddesses in Roman mythology representing the moon, which explains why in Romance languages the article for the moon has traditionally been feminine (*la lune* in French, *la luna* in Spanish and Italian, *a lua* in Portuguese, and so on). In contrast, the Germanic peoples characterized the moon—*der Mond* in modern German—as a god, Mani.

A few centuries before the Christian era, Greek philosophers took a greater interest in the moon. In fact, as Dana Mackenzie has remarked, it became "a very important test case for philosophers and their competing cosmologies." The astronomer Anaxagoras (500–428 B.C.) was the first Greek philosopher to explain the origin of solar eclipses. He thought that the sun was a red-hot stone and the moon "a stone star" made of earth, and was banished from Athens for this impious belief. The philosopher Democritus (ca. 460–370 B.C.) not only believed that there are many worlds and that all matter is composed of atoms but also proposed the idea that the lunar markings are caused by valleys and mountains. For Aristotle (384–322 B.C.) the moon marked an important cosmic boundary; he envisioned a distinction between the sublunar and supralunar worlds divided by a spherical shell in which the moon is located. According to this concept, the sphere extending from the Earth to the moon was made up of the four elements and subject to generation and corruption—in other words, it was characterized by birth, death, and change of various kinds. In contrast, the spheres above the moon are marked by the regular movements of stars, sun, and planets moving around the Earth, following the eternal laws of the divine order. They are fixed, and their motion is circular and perfect. Aristotle also deduced that the moon must be a sphere which always shows the same face to the Earth and that the moon must be nearer to Earth than Mars, because this planet is sometimes hidden by the moon.

Aristarchus of Samos (320–230 B.C.) pro-

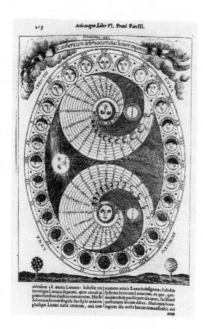

circulum 18. aetatis Lunariae: habebis tot horologia Lunaria septem, quot circulis praecedentibus diurnis noscuntur. Mis fic delineatis horum fingulis fuo ftylo menium plusque Lunari circa centrum, una cum numero aetatis Lunaris designata; habebis fyftema horarum Lunarium, ex quo, que modi modum pucib uere dicamus, facillimo nocturnam horam dicas; illud enim horologium illa nocte horam demonstrabit. cui

"The Selenic Shadowdial or the Process of the Lunation," from Athanasius Kircher's *Ars Magna Lucis et Umbrae* (1646), which gives the moon twenty-eight phases

posed—eighteen centuries before Copernicus—a heliocentric theory of the solar system and undertook serious efforts to measure the distances between Earth, moon, and sun with the help of geometry. Given the imperfect nature of his operations, most of his results deviate considerably from what is known today, but he established the relative distance to the moon with only a minor error. His estimate that about sixty Earth radii separate the moon and the Earth is within the range of the moon's elliptical orbit: the distance varies from fifty-five to sixty-three radii. (It was Hipparchus [190–120 B.C.] whose additional observations determined that the moon's motion could best be described as an oval rather than a circle.) But Aristarchus's model was far superseded as the dominant one during the centuries to come: the geocentric cosmology of the Greek Ptolemy (A.D. 90–168)—in which the moon was Earth's closest neighbor and, like the sun, orbited Earth—remained the most widely accepted view. Greek observers were probably the first to come up with the idea that the dark patches on the lunar surface were seas and the bright regions land. Somewhat later, the Latin term *mare* (plural *maria*) came to designate the dark areas, and *terrae* was used for the highlands.

In *De Facie in Orbe Lunae* (On the face in the orb of the moon), a rich dialogue and catalogue of competing ideas about the moon, the Greek biographer Plutarch (ca. A.D. 46–120) wrote that the Arcadians of pre-Hellenic Greece imagined themselves as *Proselenes,* or people whose origin reaches back to the time "before the Moon," when the moon was not yet in the Earth's sky. In claiming this history, the Arcadians attributed a special sense of nobility to themselves. The Mozca (or Muysca) Indians of the Bogotá Highlands in the eastern Cordilleras of Colombia also relate some of their tribal reminiscences to the time before there was a moon. But these myths linking human history to the moon are exceptions. For most the moon was always there, and always would be. But sometimes, the moon's position was believed to be subject to change. With the help of magic one could make the moon descend from the sky. Aristophanes (ca. 446–ca. 386 B.C.) speculated in his play *The Clouds:* "What if I were to hire a sorceress from Thessaly, and if I made the Moon sink out of the night; if I then enclosed it in a round frame, like a mirror; and then if I kept it closely guarded?"

Working during what the Western world calls the Middle Ages, Arab astronomers such as Abu Abdullah Al-Battani (858–929) should be recognized for improving many of the calculations of the classic astronomers, including that of the moon's orbit. Arabian astronomers also invented the astrolabe, which helped to perform measurements in the sky before the telescope became available. Al-Hasan Ibn al-Haytham (965–1039), who is best known for discovering the mechanism of how light enters our eyes and enables us to see, also commented on the constancy of the lunar markings and the character of the moon reflecting sunlight.

How was the distance between moon and Earth imagined in this era? Three-quarters of a millennium ago, Roger Bacon (ca. 1214–1294), the legendary English Franciscan monk and one of the first modern scientists, calculated that if a person walks twenty miles a day, he would reach the moon in fourteen years, seven months, twenty-nine days and some hours. As we know now, this space hiker would have needed more than twice as long, but who is to blame Bacon? He used the best data at his disposal.

According to the teachings of the Roman Catholic Church, based on both the book of Genesis and the works of Aristotle, the cosmos was divided into heavenly and earthly spheres. The moon—"the lesser light to rule the night" (Genesis)—which is chronicled to have come into existence on the fourth day of creation, belonged to the heavenly realm and was thus divine and composed of ether, the fifth element, the quintessence of everything. Within Christian belief systems, ancient myths connected with natural appearances often survived under different forms. For the Lusitanians, for example, an ancient people living in the western Iberian Penin-

sula before and during the time of the Roman Empire, the moon was a goddess responsible for fecundity in the human, vegetable, and animal worlds. With the advent of Christianity an important part of the lunar symbolism of this pre-Christian goddess passed to Mary. Although the ancient faith was maintained, the external appearance followed the requirements of the new times, and in the wake of this transformation the influence of the moon goddess was kept alive. The association of Mary with the moon had far-reaching consequences—like the Mother of God, the moon could not be the source of light herself, and she had to be immaculate. Although the first requirement was clearly confirmed as astronomical science progressed, the second became a challenge to the Christian worldview when Galileo

Madonna and child by Albrecht Dürer

demonstrated that the lunar surface was not in fact smooth, but rough and uneven. But this discovery was not the death knell for the connection between Mary and the moon, as the rich legacy of such images in modern times testifies.

The Catholic Church also undercut the moon's long-standing influence on reckoning of time in months and years. When Pope Gregory XIII in 1582 established the calendar that was to take his name and eventually gain international acceptance, the lunar calendar, seen by the church as profane, bowed to a system based on the sun, which symbolized the risen Christ. The Gregorian calendar replaced the Julian, credited to Julius Caesar. Before Caesar, the start and finish of the month had been determined by the phase of the moon. The Roman dictator's hybrid "lunisolar" system counted the days of the year on the basis of both moon and sun and regularized the leap year to keep the nominal and seasonal years in sync.

But how did the moon's role in early timekeeping systems develop in the first place? Even if it is hard for us to imagine, early humans many not have assumed a continuity between the crescent moon disappearing in the morning sky and the body appearing in the evening sky some days later. Over time they must have begun to understand the moon as an entity of light that always changed shape in the same way—a continuity that resulted in a recognizable pattern and, perhaps, a story.

Stone structures and megaliths grouped in

circles seem to have been connected with the sky, because their alignments can be associated with the direction in which particularly visible celestial bodies rose and set. Stonehenge, a megalithic structure in the south of England, built over a time span of a thousand years, is the best known representative of such architecture. Some astronomers have speculated that Stonehenge served as an observatory to the people who erected it and was used to predict lunar eclipses. They may have understood these occurrences as affirmations of a cosmic order, not necessarily as omens of impending disaster. There are many other such monuments. In a quest to explain the multiple rows of posts at Carnac in Brittany, the archaeoastronomer Alexander Thom showed in the early 1970s how the sight lines of various tall upright stones (menhirs) and burial mounds correspond with extreme positions of the rising and setting moon.

The erection of such structures must have been very costly, suggesting that they may have had social and perhaps economic significance beyond the religious one. Humans have inevitably thought about the cyclical nature of life on Earth: recurring periods of heat and cold, drought and rain, the life stages of the animals on which they relied, the seeds and grain they gathered, and, above all, their own living and dying. At some point it became clear that the moon's peregrinations provided a means of structuring time. Once the regular disappearance and reappearance of the moon was understood, people could count the days and group them into months and smaller periods related to the phases. Martin F. Nilsson (1874–1967), in his seminal study *Primitive Time-Reckoning* (1918), reminds us that time was first counted only when the moon was visible, omitting days when the moon was dark. In ancient Greece, for example, the day of the new moon marked the start of the month and, just like the day of the full moon, it was considered a holiday.

In some instances the moon's position in the sky was taken into account in telling time, a fact reflected in sayings among central African people such as "the moon falls upon the forest," which means that it sits low on the horizon, or "it sleeps in the open air"—is present in the sky at daybreak. We also have to remember that the time between the quarter phases is a bit more than seven days—a week in the modern sense. Nilsson proposed that the moon easily fulfilled the requirements of a timekeeping device because its monthly cycle is short enough that each night can easily be distinguished by the shape of the moon and its position in the sky at sunrise and sunset. Alas, with the exception of pregnancy, the principle of counting time in moon months has not prevailed, since there are, as Nilsson has noted, "too many months in the course of one human life."

In ancient Rome, the month was likewise divided according to the lunar phases, with the first day of the month called *kalendae* (from *calere,* to be warm or glow). Farther back, the ancient Greeks developed the Metonic cycle in an attempt to determine the correspondence between the lunar phases and particular days of the year. Meton, a mathematician and astronomer who lived in the fifth century B.C., calculated that nineteen solar years correlate to 235 lunar months or 6,940 days. After nineteen years, the same lunar phase returns on the same date of the solar year. Just once in 312 years is this cycle off for one day.

Astronomers of ancient Egypt tried to reconcile the timekeeping systems based on the stars, moon, and sun. At some point, around 3000 B.C., they defined the year as having 365 days and the months 30 days, which were further divided into periods or long "weeks" of ten days. Such a system is practical, especially in the context of bookkeeping, and Copernicus used the Egyptian year for his calculations. But because it doesn't reflect the development of the visible moon, conflicts with religious-minded people arose.

Centuries later, the Babylonians used the term *shabbatum* for the day of the full moon, and the exiled Jews in Babylon probably adopted the word for their Shabbat or Sabbath. As Peter Watson has noted, "the Sabbath was originally the unlucky day dedicated to the malign god Kewan or Saturn, when it was undesirable to do any kind of work." The Jewish Passover, in turn, falls on the day of the full moon following the vernal equinox (March 21).

In the Christian tradition the moon is used to set the date of Easter, the commemoration of Christ's resurrection. Western Christians celebrate Easter on the first Sunday after the first full moon following the vernal equinox, which can fall from March 22 to April 25. The death of Jesus, his descent to hell, and his resurrection on the third day—a Sunday—is traditionally associated with the setting and rising of the sun, but some see the older imagery of the rising moon lurking behind it. As the theologian Saint Augustine (354–430) has written: "The Moon is born every month, increases, is perfected, diminishes, is consumed, is renewed. As in the Moon every month, so in resurrection once for all time."

In parts of Asia the moon is deeply connected with regularly occurring festivals. The Kumbh Mela, which involves tens of millions of pilgrims, takes place every twelve years at four different locations in India. The climax occurs on the night of the full moon, when the assembled crowd bathes in holy rivers, such as the Ganges and Yamuna. In China and among scattered Chinese communities around the world, the fifteenth moon day of the eighth Chinese lunar month brings the moon festival or midautumn day, a traditional holiday when families gather to eat moon cakes—pastries with a filling of lotus seeds—and watch the moon. If families are separated, they at least know that their kinspeople are gazing at the same moon. Legend has it that Chang'e, the goddess of the moon and wife of the hero who shot the sun, flew to the moon and has lived there ever since. During the festival her devotees hope to catch a glimpse of her dancing on the moon.

圜桃杏
餅月秋中

FORTUNE BAKERY
SAN FRANCISCO
MOON CAKE

Dividing the seasons with the help of the moon is a cross-cultural phenomenon. The Hopi Indians, who, unlike many Native American groups, resisted influence by missionaries and settlers until the 1870s, clinging to their traditional ways, provide a useful non-Western example. According to a careful reconstruction by Stephen C. McCluskey, the Hopi based their calendar on observations of the respective positions of the sun and the moon, and their ceremonies were closely related to the agricultural cycles. In contrast to the science-based astronomy of the Greeks and Babylonians, the Hopi system was

Chang'e, the goddess of the moon, from a box for Chinese moon cakes directed toward practical ends, serving "to mark time for planting and harvesting crops or indirectly as it regulates religious festivals explicitly concerned with the success of crops." In fact, the Hopi used various names for the new moon at different times of the year: moisture moon (*Pa-muya*), big feast moon (*Nashan-muya*), basket moon (*Tühóosh-muya*), harvest moon (*Angok-muya*), and cactus blossom moon (*Isu-muya*).

The prophet Mohammed is said to have declared the moon to be the true timekeeper and its calendar the only valid one. The Islamic calendar therefore follows the lunar year (six months with 30 days and six months with 29 days, for a total of 354 days), which explains why Ramadan, the traditional month of fasting and spiritual cleansing during the ninth month of the Islamic calendar, can occur at different seasons. According to a centuries-old rule, Ramadan begins as soon as the sliver of the moon is spotted, a signal that can be difficult to spot when clouds obstruct the view. Because not all Muslims consider it lawful to collaborate with astronomers in order to predict the beginning of Ramadan, the most important Islamic holiday doesn't start at exactly the same time in all Islamic communities.

While in the West, imprecise lunar calendars have come to be associated with superstition and backwardness and are thus out of fashion today, the moon is still honored here in the calendars, if only in the word *Monday* and its equivalent in other languages. The name goes back to the Latin *dies lunae,* which conquered German tribes translated directly into their languages. In Japan and Korea the first day of the week is also associated with the moon. According to the ancient geocentric worldview, all objects moving in the sky, as opposed to the fixed stars, were called "planets." They included the sun, the moon, Mars, Mercury, Jupiter, Venus, and Saturn, and each came to stand for one particular day of the week. The moon may have declined in timekeeping power, but Monday reminds us of this older order.

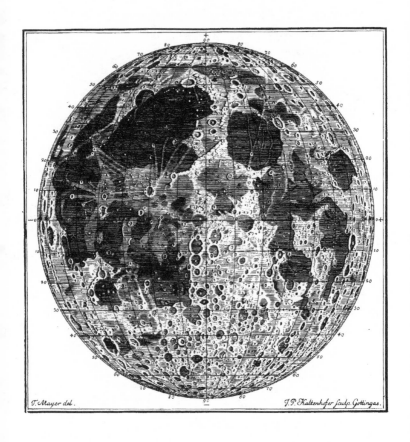

T. Mayer del.

J. P. Kaltenhofer sculp. Gottingæ.

CHAPTER THREE

Charting the Moonscape

We have the ability to imagine places and spaces that don't even exist, but sometimes reality surpasses the imagination. The discovery of America not only was a complete surprise, it also opened up the world in geographical terms and brought about a complete change in perspective. In the case of the moon the starting point was different. Its existence had long been evident, but the first glimpse of the moon through a telescope—just over one hundred years after Columbus set foot on soil in the New World—triggered a dramatic conceptual shift. No longer a mythic figure, the moon became an object whose surface could be studied in detail. Although the physical distance between Earth and moon remained the same, the moon did not seem as out of reach anymore. Suddenly an extension of one of the human senses created an illusion of much greater proximity. This was a revolution. Humankind's progressive ability to explore the moon visually surely paved the way for exploration of the moon up close with the other senses.

Just a couple of years before the telescope came into use, around the year 1600, William Gilbert (1540–1603), physician to Queen Elizabeth I, produced a rather detailed pen-and-ink sketch

Tobias Mayer's lunar map

Moon map by
William Gilbert

of the moon. This would not be noteworthy in itself, but Gilbert also introduced the first brief nomenclature for features of the moon. Gilbert, who assumed the dark spots to be land rather than seas, coined thirteen terms. What is called Mare Crisium today was "Brittannia" (Britain) for him, and his "Regio Magna Orientalis" (large eastern region) corresponds relatively well to the current Mare Imbrium. "Regio Magna Occidentalis" is a conglomerate of what are now Mares Serenitatis, Tranquillitatis, and Foecunditatis. "Continens Meridionalis" and "Insula Longa" are segments of the current Oceanus Procellarum. Gilbert's names never came to be widely used, certainly in part because they weren't published until 1651, by which time two other major schemes had been made public.

A Dutch spectacle maker, Hans Lippershey, introduced the earliest type of telescope, using two lenses, in 1608. Soon, unseen worlds appeared, but the new device first had to overcome considerable prejudice. As the art critic Martin Kemp reminds us, "strange things were becoming visible for which no ready frame of interpretative seeing existed," and the telescope "was vulnerable to the charge that what was seen through it was in whole or in part produced by the instrument itself."

Although Galileo is commonly believed to

have been the first to examine the moon with the help of a telescope, the English scientist and mathematician Thomas Harriot (ca. 1560–ca. 1621) drew the first known sketch of the moon based on a telescope viewing. Harriot made his sketch on July 26, 1609, four months before

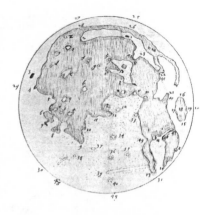

Thomas Harriot's map of the moon

Galileo's observations, and created a map about two years later. He used letters and numbers to help identify certain spots. His endeavors seem to have been known to some scholars in his field, but his renown was limited because he did not publish his lunar drawings.

Galileo Galilei (1564–1642) soon heard about the groundbreaking invention and set about improving upon it. Directing his telescope toward the Adriatic Sea, he was able to discern—as somewhat blurry images fringed with color—ships some hours earlier than with the naked eye. When he steered his device toward the moon on the clear night of November 30, 1609, he saw that its surface was not "uniformly smooth and perfectly spherical, as countless philosophers have claimed about it and other celestial bodies, but rather, uneven, rough, and full of sunken and raised areas like the valleys and mountains that cover the Earth." He wrote that the large dark spots "are not seen to be at all similarly broken, or full of depressions and prominences, but rather to be even

Galileo Galilei presents his telescope to the Doge Leonardo Donato and the Senate of Venice.

and uniform; for only here and there some spaces, rather brighter than the rest, crop up."

In his book *Sidereus Nuncius* (Starry messenger), published in March 1610, Galileo showed that characteristics of the Earth were not unique in the universe and that the celestial bodies didn't display the absolute perfection older traditions of thought ascribed to them. But despite the noted similarities between the two orbs, Galileo didn't share the view of his contemporaries that the moon was just another Earth composed of soil

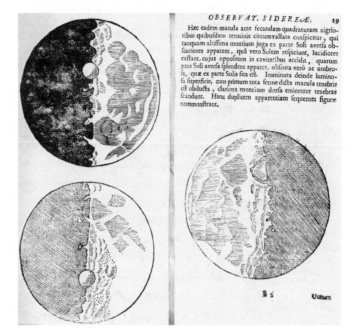

OBSERVAT. SIDEREÆ. 19

Hæc eadem macula ante secundam quadraturam nigrio-
ribus quibusdam terminis circumvallata conspicitur, qui
tanquam altissima montium juga ex parte Soli aversa ob-
scuriores apparent, quâ vero Solem respiciunt, lucidiores
exstant, cujus oppositum in cavitatibus accidit, quarum
pars Soli aversa splendens apparet, obscura verò ac umbro-
sa, quæ ex parte Solis sita est. Imminuta deinde lumino-
sa superficie, cum primum tota ferme dicta macula tenebris
est obducta, clariora montium dorsa eminenter tenebras
scandunt. Hanc duplicem apparentiam sequentes figuræ
commonstrant.

and water. He conceded that other factors than land and water might account for the differences of brightness. Galileo's great accomplishment was to significantly narrow the gap between speculation and knowledge, to bring the moon almost within grasp. He was able to see what others had not seen, but the capacity of his telescope was limited. He saw neither mountains nor valleys but, as Scott L. Montgomery has remarked, "a jagged outer edge, evolving and irregular patterns of light and shadow." In his paintings, then, he has altered various aspects for effect: "The terminator is far more irregular than in reality, and the craters are enlarged to almost double their size."

Fantasies are often more attractive than reality,

Galileo's images from *Sidereus Nuncius*

and close examination of a beloved object can reveal things we'd rather had remained hidden. The power of the telescope soon made it apparent that other satellites existed; the uniqueness of Earth was put into question. In fact, the Chinese astronomer Gan De, who lived in the fourth century B.C., had discerned Jupiter's satellites with the naked eye, but in the West Galileo is credited with this discovery in 1609. He called them *planetae*. In the following year he discovered the largest four satellites of Jupiter, namely Callisto, Io, Europa, and Ganymede, later called the Galilean moons.

Soon the development of increasingly powerful telescopes led to fierce competition to create the most accurate map of the moon—which led, ironically, to further inaccuracies. To begin with, drawing an image seen in the telescope was a challenge, requiring frequent repositioning to compensate for the movement of the moon. Competing mapmakers in different countries published their works at different stages of the production process, and on some copied maps the same feature on the surface might be perceived differently and given different names.

After Galileo, a line of scientists produced or commissioned engravings on the lunar theme. The Parisian mathematics professor Pierre Gassendi (1591–1655) provided names for some features revealed with the help of a telescope, such as *umbilicus lunaris* (the navel of the moon, now called the Tycho and ray system). Gassendi's beautiful map was executed by Claud Mellan and published in 1637.

Michael van Langren (1600–1675), also called Langrenus, came from a well-known Flemish globe and mapmaking family. His moon project was driven by the problem of determining longitude to orient ships on the oceans. He came up with the idea of timing sunrise and sunsets on "islands and the peaks of mountains most often isolated from the main continuum which in an instant appear on the face of the waxing Moon, and also those which suddenly vanish on the waning Moon, which first instant and duration is a help in finding longitude."

Mapping, of course, involves more than just creating an accurate visual representation. The identified bodies on a map need names that will resonate with those who use the map. Some artists and users were more able than others to enforce nomenclature. King Philip IV of Spain, for example, demanded that van Langren's map, which was published in 1645 with a moon measuring more than thirteen inches across, feature large formations named after the Catholic king and both living and deceased members of the royal family, such as the Oceanus Philippicus, now the Oceanus Procellarum. The royal hoped this act of naming would enhance his importance and even endow him with immortality. Scientific nomenclature played only a limited role in van Langren's map, and use of contemporaries' names for so many features partly explains why the map could not stand the test of time. Van Langren also used the names of thirteen saints, gave "water" features such names as Mare Venetum (Venetian Sea) or Portus Gallicus (French harbor), and dubbed the main highlands in honor of human virtues, resulting in features named Terra Pacis and Terra Dignitatis. Scott Montgomery recognizes the effort "to claim the Moon as Catholic territory." "The lunar surface, according to this scheme, did not belong to astronomy; astronomy, however, belonged entirely under royal power." None of the place names mentioned above survives today. On the other hand, vanity inspired van Langren to name a prominent crater Langrenus, and this name has stuck—a charming reminder of these early mapping efforts. Both Gassendi and van Langren used Caspian Sea—the name of a sea here on Earth—to designate the large dark oval spot now known as Mare Crisium. Ewen A. Whitaker assumes that this feature got its name "because it occupies roughly the same position with respect to the Moon's face that the Caspian Sea does with respect to a map of Europe, N. Africa, and the Middle East."

It is not clear whether the German-Polish politician and astronomer Johannes Hevelius (1611–1687) was familiar with van Langren's map. The hills, mountains, and crater rims in Hevelius's *Selenographia* (1647) resemble rows of termite hills, which was the usual geographic practice at this time for terrestrial maps. After considering the option of naming the lunar features for mathematicians credited with advances in astronomical science, Hevelius chose instead a profusion of geographical terms for newly discovered lunar features. As he put it: "I found to my perfect delight that a certain part of the terrestrial globe and the places indicated therein are very comparable with the visible face of the Moon and its regions, and therefore names could be transferred from here to there with no trouble and most conveniently; namely, think of the part of Europe,

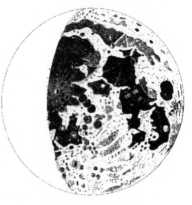

Asia, and Africa that surround the Mediterranean Sea, Black Sea, and Caspian Sea." His concept did not stand up. The names he chose were too awkward and sounded too archaic, and only ten have survived to modern times. Furthermore, his map created confusion because he accorded several groups of craters a single name and saw mountain features where none existed. Elsewhere, he mistook peaks for craters. Still, despite its shortcomings, Hevelius's map retained its primacy for almost one and a half centuries.

Drawing of the moon by Johannes Hevelius

The Italian Giovanni Battista Riccioli (1598–1671), who competed with Hevelius, used 63 of van Langren's names in his 1651 *Almagestum novum* (New almagest), but applied three of them to different lunar features and added 147 new names taken from persons connected to astronomy. Doing away with the catalogue of moral qualities, he chose names related to the weather on Earth, such as Terra Caloris (land of heat) and Terra Nivium (land of snows), none of which are used on maps of today. Riccioli was convinced that there were no humans and no water on the moon, and, as a Jesuit, he was obliged to reject the Copernican system—though it is revealing that he named three prominent craters after Copernicus, Kepler, and Aristarchus. The basic nomenclature devised by Riccioli—assigning the names of famous scientists to craters and classical Latin names characterizing weather or states of mind to maria (such as Mare Crisium and Mare Serenitatis)—has survived to the present day, though the names themselves have changed.

Beginning in 1748, the German astronomer Tobias Mayer (1723–1762) made more than forty detailed drawings of various lunar regions in an attempt to create a more accurate map. Although the moon globe he envisioned was never constructed, two lunar maps and some drawings eventually emerged. The engraver reversed one drawing by mistake, so the plate had to be accompanied by the following caption: "If you want to view the print correctly, you must hold it up to a mirror."

Johann Hieronymus Schröter (1745–1816) based his map on the measurements Mayer had performed. He surpassed his contemporaries in not only creating a vast number of drawings of many smaller areas of the moon at various states of illumination but also approximating both mountain heights and crater depths by measuring shadow lengths. Since there was now a consensus that water could not exist on the moon's surface, Schröter had to face the problem of outdated terms like *peninsula, river,* and *swamp* in

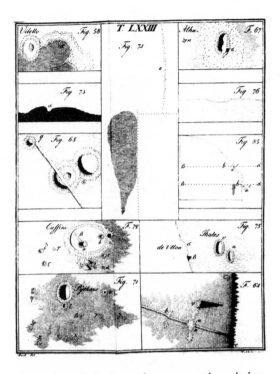

older maps. As Latin no longer was the sole language of scholarship, he also introduced the German word *Rille* for "groove."

The most influential contribution to lunar topography in the nineteenth century—a highly detailed large-scale lithographic map that surpassed all predecessors in refining the system of feature names and positions based on micrometric measurements—was created in Berlin by Johann Heinrich von Mädler (1794–1874) and his friend Wilhelm Beer (1797–1850). Beer was a banker who provided the observatory and a four-inch refracting telescope and Mädler was the observer, artist, and scientist who drew the map.

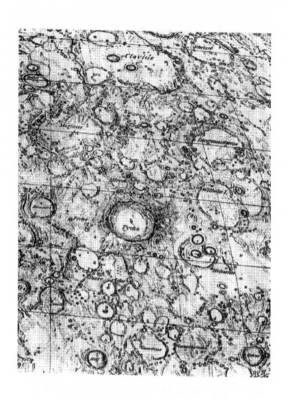

A detail from the map by Mädler and Beer

Studying the moon for about six hundred nights, Mädler created a *Mappa Selenographica* more than three feet square that depicted the moon not in spatial lunar coordinates but as a disk with contortions on its fringes. He designated secondary craters by assigning each a letter associated with a nearby "parent" crater, and he assigned Greek letters to elevations, ridges, and rilles. Mädler concluded that the moon has no atmosphere or water and does not change: "The Moon is no copy of the Earth." Part of the publication was a bulky volume titled *Der Mond,* which was a resource of knowledge about the moon, "our satel-

lite which in front of the eyes of all inhabitants of the Earth repeats its eternal cycle."

Johann Heinrich von Mädler and a moon globe

In 1840 Mädler became the director of the Dorpat Observatory in Estonia. Based on the work with a larger refracting telescope available there, he published a revised version of his map in 1869. By this time, photography was starting to decrease the interest in lunar cartography, but illustration had the advantage of allowing emphasis of detail. The revised edition of Mädler's book seemed to answer conclusively the most important questions about the physical condition of the moon; the book had the effect of effectively paralyzing the study of the moon for decades.

As lunar maps evolved, their functions became

more firmly fixed. Certainly they helped to serve the vanity of rulers who commissioned them, but otherwise they served no obvious political function, such as the territorial claims that terrestrial maps articulate. And unlike maps of Earth, visual representations of the moon didn't help travelers orient themselves in little-known terrain. Their function was largely symbolic—even though the moon was out of physical reach, mapping it had an imaginative value. More than at any previous era, the nineteenth century was obsessed with counting and measuring—an activity that engaged the many blank spaces and uncertainties related to the cosmos without quite solving them. Precise maps also had the effect of elevating science at the expense of ancient myth.

The degree of concentration required to chart a world so utterly out of reach was immense, and before the advent of photography it was a tedious task to observe the details and then transfer these impressions with a pencil. Mädler's obsession with his project was such that he is said to have met his wife in one of the rare moments when he was not looking through the telescope. The German astronomer Johann Friedrich Julius Schmidt (1824–1884) is credited with the most detailed lunar map of the nineteenth century, featuring no fewer than 32,856 craters. Such an obsession with ever-smaller features, as Paul D. Spudis has stressed, was inextricably linked with the idea common during this time "that the secrets of lunar history were in the details" rather than in the broader features.

The invention of photography had an impact on the depiction of the moon, just as it did on all forms of visual representation. In 1840 John William Draper (1811–1882), then a professor of medicine at New York University, created a few crude daguerreotypes of the moon. They were soon superseded in quality by the photographs by his son Henry (1837–1882), and those of William Cranch Bond (1789–1859), the first director of the Harvard College Observatory, who used a refracting telescope of fifteen inches, then the largest in the world. A particular challenge facing photographers was the need for long exposure times combined with the movement of the moon in the sky, which led to blurred images. But the complication could be resolved. A clockwork mechanism was attached to the camera to compensate for the rotation of the Earth and moon. Another pioneer of lunar photography was Lewis Morris Rutherford (1816–1892). A lawyer by education, he started to devote himself to astronomical photography in the 1850s and soon maintained a little observatory in Manhattan. Rutherford also designed a special telescope for this kind of photography. The quality of photographic emulsions steadily improved, but well into the twentieth century scientists continued to rely more on illustrations of the moon, which were richer in detail, than on photographs.

How did the progressive discovery of the moon fit into the larger development of astronomy? The observation and mapping of the moon in the centuries after Galileo continued to be-

PHOTOGRAPH OF THE MOON.

Photograph taken by Henry Draper in 1863

come more precise, and astronomers grew ever more interested in the planets. Some of the scientists we have met in this chapter also began to study Mars and other planets. In 1830, for example, Wilhelm Beer published the first map of Mars. In 1869 Richard Proctor produced an even more detailed map. Cassini also devoted much of his attention to the Red Planet.

By the end of the nineteenth century, the telescope, now much improved and widely used, had rendered the moon rather mundane. Many researchers now devoted their energies to more promising fields of stellar enquiry. Some lamented that the profession had abandoned the moon to amateurs. What attention it did receive was mostly in the context of its surface and of larger issues of space exploration. But in the 1960s, some geologists looked up from Earth and began to take an interest in the surface and makeup of the moon. They began also to transform their methods to accommodate lunar exploration.

The scientist connected most with this major transition is the American geologist Eugene Shoemaker (1928–1997), who had devoted himself to the study of cosmic impacts, examining Barringer Meteor crater in Arizona and craters blasted by atomic bombs tested in Nevada. Realizing that geologists would eventually be needed to investigate the composition of the moon, Shoemaker founded the astrogeological branch of the U.S. Geological Survey in 1961, and the group began using telescopes and photographs to map lunar geological features. One result of this work, a few years before the moon could actually be scrutinized and mapped up close, was the *Photographic Lunar Atlas* (1960, 1967), published by the U.S. Air Force and the University of Arizona, which depicts the moon's near side under various light conditions. Originally a potential Apollo astronaut, Shoemaker had to abstain from a lunar trip for health reasons. But in another sense he still made it to the moon—his cremated remains were deposited there by the mission Lunar Prospector in the year after his death. So far, he is the first and only human interred on the moon.

Lunar features continued to be named and re-named throughout the twentieth century. Lunar nomenclature is a discipline both esoteric and complex, often resulting in confusion, if not chaos. Ewen A. Whitaker's *Mapping and Naming the Moon* (1999) deals with some of these "nomenclatural nightmares." When new features were discovered—for instance, after the photographs were made of the moon's far side—the lack of specific rules became apparent. Special committees were formed and international resolutions adopted. Thus it was determined that craters can be named only after deceased people and that specific types of lunar features, such as mountains, rilles, and valleys, must named in Latin: mons, rima, vallis.

Over time, improved resolution of images— as with the metric camera photography from Apollo missions 15, 16, and 17—brought new challenges in nomenclature: In how much detail should maps represent the surface now that even much smaller formations became discernible? And while few of the fanciful names given centuries ago to lunar features have survived, the dark patches continue to be called maria. Lunar nomenclature continues to reflect the advancement of scientific knowledge—with some erratic leftovers from older times.

Visual exploration of the moon only foreshadowed a much more complex scientific enterprise. Twenty-first-century study of the moon extends well beyond sight. Spectrometers allow for ever more precise analyses of the moon's mineralogy and composition. Measurement of gravitational fields, temperature, and radiation are becoming increasingly accurate. Navigation data of lunar spacecrafts revealed the existence of mascons: patches on the surface, mainly on the near side, with particularly high gravity. Efficient camera systems are able to detect the details of rock formations hardly three feet high. The Japanese lunar explorer Kaguya (Selene) has, since 2007, circled the surface at a distance of roughly sixty miles, using so-called laser altimeters that constantly send impulses to provide highly detailed three-dimensional information about the moon's topography.

These efforts to map the moon and name its features reveal that conquest no longer is the exclusive province of kings and gunships, but includes detailed scientific investigation. The moon is more than an object of study and contemplation; it is also a screen upon which our desires and aspirations can be projected. It is fascinating to note that the Italian philosopher and optician Giambattista della Porta (ca. 1535–1615) took this idea literally and proposed precisely such a project. In his book *Magia naturalis* (1589), he suggested that the moon could become a news medium. According to his proposal, a parabolic mirror with a wide focus would project letters onto the lunar surface that could then be read by people on Earth. Della Porta's idea was never put into practice, and the stories that shake the world today are to be found on computer screens, not on a lunar screen.

CHAPTER FOUR

Pale Sun of the Night

That the moon shines is irrefutable, but just how bright is it? There is no easy answer, because the intensity of moonlight—measured in lux—varies considerably and is influenced by several factors. During a full moon the luminance is about 25 times as great as at the time of the quarter moon and 250 times as great as on a clear, moonless night. The distance between Earth, moon, and sun, which changes because the terrestrial and lunar orbits are not perfect circles, also has a bearing on the intensity of light. The moon's lighter lunar highlands reflect much more sunlight than do the darker areas—12 to 18 percent compared with 5 to 10 percent. Also affecting the moon's brightness is the phenomenon of atmospheric extinction—that is, the loss of light as it passes through the Earth's atmosphere. In dry, clear air, the atmosphere has relatively little impact, but moist or dusty air mutes the intensity of moonlight; thus moonshine can appear particularly intense in dryer climates. Apart from the air quality, the very amount of air through which the light beams pass—measured in units of "air mass"—can vary. Moonlight passes through one air mass when the moon is directly above us, forty when it is on the horizon. These two fac-

Moonlight as atmosphere

tors can combine to produce enormous variations in how bright the moon appears. But even at its most intense, the moon's light is of dramatically low intensity compared with the sun.

But a question remains. A piece of jet-black coal exposed to sunlight here on Earth is black no matter what, so why do we perceive the full moon in the sky as white even though the material that makes up its surface is gray or black? The standard explanation used to be that the diffuse reflection of all the wavelengths of sunlight causes the perception, but recently scientists from different disciplines have suggested a different hypothetical source for the illusion. According to Sobha Sivaprasad and George M. Saleh of the Princess Royal University Hospital in Orpington, Kent, "the eye computes the colour based on relative inputs." Compared with its surroundings, the black of space, the moon undergoes "an artificial brightening," and as a result, we visualize the gray moon as white.

Robert W. Kentridge of the University of Durham offers an alternative explanation. He believes that the moon "is most likely seen as white because it is mistakenly perceived as being lit by the same illuminant as its surroundings when, in fact, its direct scattered illumination is far more intense than the scattered illumination of its apparent surroundings." Moon rock viewed up close on Earth appears dark "because there is no confusion about its illumination." As Paola Bressan of the University of Padua reminds us, the understanding of this mechanism can be credited to Adhémar Gelb. Gelb, in 1929, suspended a disc made of black paper in a dark room, then illuminated it with white paper. "The fact that it was actually black became evident only when a larger surface of higher luminance, such as a sheet of white paper, was brought into the beam of light and placed behind the black disc," Bressan explains. "But as soon as the white paper was taken away, the black surface went precipitously back to white, demonstrating that perception was impermeable to the knowledge of the 'real' colour of the disc." The so-called Gelb effect shows "that the highest luminance in a scene appears white" and confirms that "the moon looks white simply because it is the brightest region in the nocturnal sky."

Our perception of how sunlight and moonlight affect the world around us is a study in opposites. The intensity of sunlight makes everything appear etched in sharp contours and gleaming in its brightest colors. Sunlight is a metaphor for clarity and truth, making misperception all but impossible. Against the dark sky, the moon appears even whiter than the sun, but its light seems to drain the world around us of its color and veil everything in shades of gray. Nearby objects appear far away and vice versa, tricking us into believing what is untrue. By the light of the full moon, objects motionless during the day become eerily animated and changeable—the dim light leaves us room to interpret the evidence of our senses.

If we consider physical properties alone,

moonlight is practically the same as sunlight, despite how different they appear to the human eye. Objects look gray by moonlight only because moonlight falls just under the threshold at which color-sensitive receptors in the retina function effectively, but well above the much higher threshold for simple light receptors. Long photographic exposures under moonlight with color film have demonstrated that the colors are in fact there, even if we are not able to discern them. If we look at a landscape illuminated by moonshine long enough for our eyes to adapt to the dark, the gray appears blue. This phenomenon of blueshift is the result of a particular response of the rods, the photoreceptor cells in the retina.

Some philosophers of ancient Greece, such as Thales of Miletus, Anaximander, Pythagoras, Parmenidis, and Empedocles, theorized that the moon shines because it reflects the light of the sun. The Egyptians had also made this discovery, but this news reached western Europe only much later, in the twelfth century. Even then, uncertainty long persisted about whether the moon generates its own light. Among his many more celebrated pursuits, Leonardo da Vinci (1452–1519) devoted some attention to this question:

> Either the Moon has or has not its own light. If it has its own light, why does it not shine without the help of the Sun? And if it has not its own light, it must be a spherical mirror. The Moon does not have its own light but shines only as long as the

Sun makes it shine; and we see exactly as much of its light as it sees of ours. Its night receives as much brightness as our waters lend to it as they reflect the image of the Sun on it, for in all the waters which they look down upon, the Sun and the Moon are reflected. The Earth and the Moon lend each other light.

Leonardo didn't have it quite right: Earth shines on the moon with much greater intensity than the other way around. In fact, this light is so bright that it is perceptible to us by means of a secondary reflection. Some days before and after a new moon, though only a crescent is illuminated by the sun, the rest of the disk is clearly visible as well—"the young moon with the old moon in her arms." It faintly marks out the night hemisphere and grows in intensity with the narrowing of the crescent or vice versa. This is the earthlight or earthshine of the moon: a reflection of a reflection or secondary moonlight, also called *clair de terre*. The light has traveled three times—from the sun to Earth, from Earth to moon, and from moon back to the eyes of the observer on Earth. When the moon is full, this shine is no longer noticeable from Earth.

The explanation of earthshine historically presented some problems. Sometimes it was attributed to the moon itself, thought to be transparent, or to the reflection of Venus. The Aristotelian notion that the moon mirrors the earth was developed further by some eighteenth- and nineteenth-century theorists. The German physicist and astronomer Johann Heinrich Lambert (1728–1777) may have taken the theory a step too far. He claimed to have observed, on February 14, 1774, that the ash color of the moonlight changed into an olive green bordering on yellow. He explained the phenomenon as follows: "The moon, which then stood vertically over the Atlantic Ocean, received upon its night side the green terrestrial light, which is reflected toward her when the sky is clear by the forest districts of South America." Alexander von Humboldt (1769–1859) was probably on solider ground when he observed that "the extremely variable intensity of the ash-gray light of the moon depends upon the greater or less degree of reflection of the sunlight which falls upon the earth, according as it is reflected from continuous continental masses, full of sandy deserts, sandy steppes, tropical forests, and barren rocky ground, or from large ocean surfaces." And Camille Flammarion (1842–1925), the great French popularizer of astronomy, wrote, "This ashy light, the reflection of a reflection, resembles a mirror in which we may see the luminous state of the Earth. In winter, when a great part of the terrestrial hemisphere is covered with snow, it is perceptibly lighter." He also cites a very curious case: "Before the geographical discovery of Australia, astronomers suspected the existence of that continent from the ashy light, which was very much brighter than could be produced by the dark reflection from the ocean."

In ancient India, the moon was called "the king of the stars of cold" and was thought to radiate coolness. Plutarch came closer to the truth, complaining "that the sunlight reflected from the moon should lose all heat, so that only feeble remains of it were transmitted by her." The moon does emit some heat, but it's hardly measurable. In 1725 the French geophysicist and astronomer Pierre Bouguer (1698–1758), the founder of the science of photometry, studied the illumination cast on a screen by the image of a full moon in a concave mirror. After repeating the experiment during four nights, he wrote, "The sun illuminates us about 300,000 times more than the moon." In 1860 the director of the Harvard University Observatory, George Bond (1825–1865), used a large glass sphere with a silvered reflecting surface to determine that the sun is 470,980 times brighter than the full moonlight. The ratio of brightness as determined by modern analytical methods is about 400,000 to 1.

In 1846 the Italian physicist Macedonio Melloni (1798–1854) tried to determine the temperature of lunar rays reaching the atmosphere during the different phases of the moon from his meteorological station high on Mount Vesuvius. The experiment was repeated ten years later by Charles Piazzi Smyth (1819–1900), the Royal Astronomer for Scotland, who ascended Mount Guajara in Tenerife armed with a state-of-the-art temperature gauge. Although Melloni could find only the smallest elevations of temperature,

Smyth compared the heat to one-third that of a wax candle placed at a distance of about sixteen feet.

In the industrialized parts of the world, the moon's luminescence is no longer what it used to be. Thousands of generations before us lived in a world with no artificial light and had a much different experience of the night sky. Back then day and night were regarded as more different than they are today—*were,* in fact, more different. If you happen to live in a large city or metropolitan area, you never experience pitch dark. The moon and some stars and planets may be visible, but they compete with a vast number of street lamps, lighted billboards, and other sources of illumination. The English essayist Alfred Alvarez goes so far as to say that if you live in a large city, you forget the night—and the moon, we might add. Only during a power blackout or when we walk in a forest at night do we have a chance to taste what it must have been like in earlier times.

How was the experience of moonlight different? In earlier times people were used to spending a great deal of time outside, where their bodies and senses experienced bright days and dark nights, and the rhythms of their bodies, including their sleep patterns, adapted accordingly. Nights without the moon in the sky had a dramatic quality. As Jérôme Carcopino wrote about ancient Rome:

> When there was no moon, its streets were plunged into impenetrable darkness. No oil

lamps lightened them, no candles were affixed to the walls; no lanterns were hung over the lintels of the doors, save on festive occasions when Rome was resplendent with exceptional illuminations to demonstrate her collective joy. . . . In normal times night fell over the city like the shadow of a great danger, diffused, sinister, and menacing. Everyone fled to his home, shut himself in, and barricaded the entrance. The shops fell silent, safety chains were drawn across behind the leaves of the doors; the shutters of the flats were closed and the pots of flowers withdrawn from the windows they had adorned.

Of course, there were fires and eventually fireplaces and—only much later—candles and oil lamps. But candlelight was introduced around the early sixteenth century and was at first popular only among privileged people. For the vast majority of people nighttime darkness was the rule, interrupted only by nights with intense moonlight. And even the early sources of artificial light produced only a fraction of the illumination of the ubiquitous electric lights of today, and their disruption of the natural rhythm of darkness and light was therefore minimal.

In a world dominated by natural illumination, the moonlight visible on a night with a clear or only partly cloudy sky took on a special significance. People stayed up longer and slept less during such a night; they lived by the moon—their pale sun of the night. Farmers often could perform chores by moonlight that they were not able to finish during the day, especially when all the fruit was suddenly ripe. In the fall, under the light of the harvest moon—the full moon rising at around sunset for several evenings in a row—they could continue to gather their crops long after the sun had set. Centuries earlier, hunters and gatherers had to find ways to make the best use of limited light; during moonlit nights they were able to extend their pursuits into the night as well. When we say today that somebody is "moonlighting," undertaking some unusual or extracurricular activity for profit, we invoke an earlier time when people fulfilled their duties after dark.

Contemporary nomadic tribes in the Sahara Desert seek protected spaces during the day when it's too hot to travel and take advantage of the moon as a gentle guardian to guide them as they journey through the cool dark. I have seen fishermen searching the shallow waters of the Red Sea by moonlight, only occasionally using an additional lamp to attract their large and colorful prey. And in our high-tech culture, some hunters take a special thrill from catching prey under adverse circumstances; special night-vision lenses aid such hunters by gathering light from the moon and stars and directing it into a photocathode tube. By converting photons to electrons, these devices increase the hunter's visual acuity.

But even if the only activity was a stroll, before the competition of artificial light, the moon

showed the wanderer the way to the well and back to his house. Old clocks mark not only the hours of the day but also the phases of the moon, so that travelers could plan their journeys for full moons, to take advantage of the extra hours of visibility. A moonlit night was never as bright as day, but the moon's bright glow could at least allow nocturnal wanderers to see the contours of the landscape in some detail. Henry David Thoreau, an astute observer of many natural phenomena, often deliberately sought refuge from civilization by rambling at night in unpopulated areas. He wrote in his journal, "I saw by the shadows cast by the inequalities of the clayey sandbank in the Deep Cut, that it was necessary to see objects by moonlight as well as sunlight, to get a complete notion of them. This bank had looked

The moon gives the icy landscape along the northern Siberian coast a particular intensity.

much more flat by day, when the light was stronger, but now the heavy shadows revealed its prominences." He was also led to think of the moon being in a "continual war with the clouds," a war that had implications for the orientation of anyone crossing the landscape at night. When the moon was absent from the sky, travelers risked losing their way. Perhaps this disorienting effect explains why Julius Caesar supposedly preferred to start his battles when the moon was dark.

Moonlight is often most spectacular in winter, when it can produce dramatic reflections in the snow. A similar effect can be observed on long stretches of beach or dunes. In the mountains, when the moon is close to the horizon, its surface resembles a faraway rocky peak, which led some early observers to conclude that the moon is made of the same matter as the Earth. Along these lines, Sir William Herschel's son Sir John Herschel (1792–1871), in his *Outlines of Astronomy* (1849), compared the illumination of the moon's surface he had observed during his trip to South Africa with that of weathered sandstone rock in full sunshine: "I have frequently compared the moon setting behind the gray perpendicular façade of the Table Mountain illuminated by the sun just risen in the opposite quarter of the horizon, when it has scarcely been distinguishable in brightness from the rock in contact with it. The sun and moon being nearly at equal altitudes and the atmosphere perfectly free from cloud or vapor, its effect is alike on both luminaries." On the other hand, moonlight can convey a remarkable effect when reflected by the surfaces of mountaintops. When the German writer Johann Wolfgang von Goethe (1749–1832) watched Mont Blanc, the highest mountain of the Alps at the border of France and Switzerland, at night, he saw "a broad radiant body belonging to a higher sphere; it was difficult to believe that it had its roots in earth." Even buildings can develop a particular magic under the moon's light. Jacqueline Kennedy Onassis is known to have returned to the Taj Mahal, the famous Indian mausoleum made of white marble, to see it again when veiled in moonlight.

When the moon is visible during the day, the quality of its light is much different. No longer even slightly yellow, it appears white under the influence of the blue light of the atmosphere, its complementary color. In *Cosmos: A Sketch of a Physical Description of the Universe* (1868), Alexander von Humboldt compared the sunlight reflected from the moon with that reflected by a white cloud during the day and found it "not infrequently difficult to distinguish the moon between the more intensely luminous masses of clouds." High in the mountains, Humboldt found that detecting the lunar disk was much easier because "in the clear mountain air, only feathery cirri are to be seen in the sky." Due to their loose texture, these clouds reflect sunlight, "and the moonlight is less weakened by its passage through the rarer strata of air."

Moonlight once facilitated social gatherings and helped slaves to find the path away from their oppressors, but on the other hand a night with the full moon in the sky encouraged thieves and criminals in their pursuits, since they didn't have to risk discovery by using artificial light. Despite a certain danger involved in crossing the landscape at night, the moon's presence makes solitude more bearable to the lonely wanderer. And if he needs to rest or sleep in the open, the moon serves as a watchlight, providing a sense of protection. Today's Zande people, living along the Nile-Congo watershed in central Africa, use moonlight ritually, sharing trickster tales during moonlit nights, woven together a little bit differently at each telling.

The moon's significance must have been immeasurable for travelers onboard ships crossing the oceans. During a tempest, when everything seemed on the verge of vanishing into nothingness, a full moon, often just lurking from between the clouds under these conditions, had the dual role of illuminating the ghastly scene and being the only beacon offering a fixed position to the people in peril. More recently, it has been speculated that had there been a full moon, the lookouts on the *Titanic* might have been able to discern the iceberg in time to change the course of the ship. But the moon's value for travelers is not only practical. Henry Matthews, who in 1817 traveled from England to Lisbon for his health, began his reflections on deck one night with an interesting comparison: "If the sunrise be best seen on shore, the moonlight has the advantage at sea. At this season of repose, the absence of *living* beings is not felt. A lovely night.—The moon, in this latitude, has a silvery brightness we never see in England.—It was a night for romance." Even today, when vast distances are usually covered by airplane, memorable moments sometimes surprise us, as anyone knows who, during a long intercontinental night flight, has been lucky enough to have seen the moon lighting the ocean surface below.

Of course, moonlight has not only served to extend typical daytime activities into the night. The array of activities for which moonlight is preferable can never be complete. The poet Rupert Brooke (1887–1915) frequently bathed in the river by moonlight with his childhood friend Virginia Woolf. Common lore across cultural barriers often associates the moon with love and other strong human emotions. "You are as beautiful as the full moon" is a common compliment in Arabic, whose alphabet has a range of lunar letters. The moon, often personified with a gentle and comforting smile, is deemed able to observe humankind from afar, possessing secret wisdom or wielding supernatural control over events. It's always up there, and it sees everything. When a loving couple secretly meets under its glow, the moon is the only other witness—perhaps the first to become aware of this bond. The lovers' awareness of the moon intensifies their emotions, and the moon becomes a recurring symbol of, even a witness to, their union. The moon assumes a similarly watchful role in many children's stories, where it sends its light through the night as it protects sleeping children. This resonates with ancient myths in which the moon acts as a judge.

In fact, as Doreen Valente has remarked in her classic account *Where Witchcraft Lives* (1962), it generally acknowledged "that moonlight has a sexually stimulating and exciting effect upon human beings." But it is difficult—sometimes impossible—to distinguish the physical impact the moon has on us from the impact of our cultural associations. If the circumstances are right, moonlight may induce a romantic or sexual mood. And because this mood immediately calls forth certain contexts, memories may be evoked that give rise to such associations more or less automatically. Making love under a full moon in an open sky has an honored place right next to the candlelight rendezvous. Yet for all its romantic power, such a scene could have prosaically unpleasant consequences. Intense moonlight, by overwhelming the senses via sheer beauty, exalted feelings, or the stimulating company of an adored person, can make those experiencing it forget the risks involved in exposing the nude or partially clad body to the elements. Marcel Proust must have had something of this sort in mind when he wrote about the "man who has forgotten the glorious nights spent by moonlight in the woods" but still suffers "from the rheumatism which he then contracted." Obviously aware of the moon's noxious potential, the French novelist also invoked its diurnal innocence, when the moon is "no more than a tiny white cloud of more definite and fixed shape than other clouds."

Finally irresistible for Proust was "the ancient unalterable splendor of a moon cruelly and mysteriously serene."

The moon has aided less amorous pursuits as well. The Lunar Society, a club of prominent businesspeople, naturalists, and intellectuals, met regularly for some fifty years starting in 1765 in Birmingham, England. Its members, including Erasmus Darwin, the grandfather of Charles Darwin, met to perform experiments and to discuss the latest developments in chemistry, electricity, medicine, and economics. The society's name commemorated the timing of its meetings, which occurred on the first Monday night after the full moon, when the additional light made the journey home safer.

The romantic moon

A particular climate of scientific inquiry and artistic exploration, combined with a Mediterranean sensibility, has made Italy a hotbed of lunar musings. The tradition stretches back at least to Dante Alighieri, an intimate scholar of celestial science, who, in his *Divine Comedy*, meets the inconstant souls in the sphere of the moon. Leonardo, Giordano Bruno (1548–1600), and Galileo carried on the tradition of lunar devotion, which culminated in the melancholic moon poems of Giacomo Leopardi (1798–1837). In Italian culture, the moon is not just something to be contemplated; it amounts to a persona with particular capabilities or an understanding being that can be addressed—a conceit carried into a modern Italian-American community in the film *Moonstruck*. As Italo Calvino (1923–1985) once wrote, "As soon as the moon has risen in the poet's verses, it has always had the power to communicate a sense of weightlessness, of suspension, of mute and calm enchantment."

But Italians scarcely have a monopoly on lunar devotion. In the English-speaking world, literature and other forms of artistic expression abound with similar patterns of emotion. William Shakespeare's *A Midsummer Night's Dream* takes place in a land of fairies under the light of the moon. Here the moon is a powerful force with an intoxicating effect on the characters, inciting bizarre and at times illicit behavior. The moon is connected to dreaming and to a world in which fantasy and reality can blur. As Duke Theseus of Athens looks forward to his marriage—and the tensions and pleasures of the wedding night—he invokes the moon as a witness that

> the like to a silver bow
> New-bent in heaven, shall behold the
> night
> Of our solemnities.

At the same time, the very moon that blesses the lovers joining in matrimony also threatens this social bond with lust and chaos. And in another play by the same author, because the inconstant moon is a threat to unending love, Juliet forbids Romeo to swear by the moon.

The moon and its light have been characterized in many ways, but often as soft, pale, sweet, enchanting, serene, lovely, mysterious, gentle, and calm. It has been seen as warm or cold irrespective of the actual temperature, but according to the accompanying psychological evocations. The moon is sometimes connected to the idea of melancholy, perhaps because of the perceived paleness of its light, perhaps because we are particularly susceptible to this state of mind at night. The English poet Percy Bysshe Shelley (1792–1822) compared the moon to a "dying lady lean and pale,/Who totters forth, wrapp'd in a gauzy veil."

In some cultures the moon stands only for innocence and purity. For example, a saying in Sri Lanka holds that "love has to be as pure as the moonlight." In other traditions, though, the moon has a dark side, both literally and metaphorically: it is a part of the night—a time apart in which the order and continuity of the daylight world are suspended. Night can provide the necessary rest from a busy day, but it is also sometimes associated with insomnia, ghosts, and even death—and so is the moon. On moonlit nights figures skulk, glowing eyes peer from bushes, and crazy laughter echoes; often the boundaries of love and death, madness and sex get confounded. The moon can symbolize darker impulses and was at times associated with immorality. Writers of the Gothic literary tradition in various cultures have exploited the fears associated with the night and the moon. In *Lord Jim,* Joseph Conrad provides a vivid example: "There is something haunting in the light of the moon; it has all the dispassionateness of a disembodied soul, and something of its inconceivable mystery." *Murder by Moonlight* is not only the title of a book and a movie but also a common artistic trope. In this genre, the moon is often counterpointed by crepuscular and nocturnal animals like owls and bats—or even vampires and werewolves, in fantasy and horror novels and films too numerous to catalogue.

But viewed from a different perspective, the moon also brings light to the darkness; it softens the difference between day and night, thus soothing fear and superstition. Paradoxically, the moon thus serves as a harbinger of hope against the very fear and superstition that the darkness often inspires. In this role, the moon can evoke Enlightenment rationalism of the kind embodied by the Lunar Society science—a view that not only rejects superstition but also promises to illuminate all subjects shrouded in darkness.

In paintings, the sight of the moon in the background or overhead lends a dramatic quality, a magical aura, to the proceedings depicted. *Flight into Egypt* (1609/1610) by Adam Elsheimer—a painter who seems to have been familiar with Galileo's work—is remarkable not only for its representation of the moon with all major maria but for the rich evocation of a nocturnal landscape with different sources of light, including the reflection of the full moon in a lake. Romantic artists have created an abundance of paintings featuring the moon. It appears close, but is in fact far away. Often, it takes on mythical

or religious symbolism—the awe of nature, for example, or the power of creation. Artists have portrayed its light reflected in lakes and oceans, peeping between the branches of trees, or shining over plains, mountains, cities, and harbors. Such paintings allow us to uncover an entire taxonomy of lunar moods. By presenting the moon's more appealing aspects, these artists helped to shape a new attitude toward the night as a time to feel at ease, a time to be savored.

Some of the dark pastoral scenes painted by Joseph Wright of Derby (1734–1797) were lit exclusively by moonlight. One example, located in Naples, is *Virgil's Tomb*. In this work the lunar light is represented with such an intense quality that it illuminates even the inner parts of a tomb. Wright also chronicled meetings of the Lunar So-

A Summer Night on the Rhine (1862) by Christian Eduard Böttcher (1818–1889)

ciety, taking the opportunity to learn Enlightenment ideas firsthand.

A modern example of the metaphorical use of the moon in literature is *Journey by Moonlight* (1937) by the Hungarian novelist Antal Szerb (1901–1945). Here the moon is a backdrop for the protagonist Mihály's feverish odyssey through Italy, where he breaks free from conformism, drifting into his dreams, his past, and his longing for death. Other artists, though, have rejected the moon's power over our imaginations. Filippo Tommaso Marinetti (1876–1944), an Italian futurist who later became an ardent follower of the dictator Benito Mussolini, expressly rejected traditional artistic concepts and ideals by exclaiming in his manifesto, "Let's murder the moonshine." For Marinetti, the moon was connected with the "sentimentality and lechery" of love, love of the "fragile woman, obsessing and fatal, whose voice, heavy with destiny, and whose dreaming tresses reach out and mingle with the foliage of forests drenched in moonshine."

Classical music, too, offers various references to moonlight. The *Suite bergamasque,* among the most famous piano compositions by Claude Debussy (1862–1918), includes a movement known as *clair de lune* (moonlight), which was probably named for the poem by the same title by Paul Verlaine (1844–1894). Gabriel Fauré (1845–1924) also composed a calm mood song carrying this title. It is a bit odd to realize that the composer of the most famous example of moon music—*Moonlight Sonata* by Ludwig van Beetho-

ven — had nothing to do with the title, which was attached to the piece after the composer's death. A music critic, while listening to the piece, was reminded of moonlight shining upon Switzerland's Lake Lucerne.

Moon music

The moon also features in much modern music. "Blue Moon," composed by Richard Rogers and Lorenz Hart, has been sung by famous singers of which there are almost too many to name, including Frank Sinatra, Billie Holiday and the Supremes. In *Tintarella di luna* (Moon tan, 1959) the Italian pop singer Mina sings about a girl who chooses to "tan" in the moonlight rather than in the sunlight so that her skin will become the color of milk. This complexion makes her "a beauty among beauties" in a land where the Mediterranean sun makes bronzed skin commonplace. She spends "all night up on the roof" like a cat. In a time before political correctness, the rays of a full moon promise to make her completely white, more white and beautiful than any other girl.

Some of the most impressive works of art relating to the moon and moonlight can be found in Japan. The moon was sacred to Tsukioka Yoshitoshi (1839–1892), and during the 1880s, he produced a formidable series of woodcut prints under the title *One Hundred Aspects of the Moon,* with figures from both Japanese and Chinese mythology at its center. In the Japanese aesthetic, the full moon is an object of beauty often linked with autumn. This season evokes melancholy because of the imminent passing of the year, in contrast to the Western mood of harvest celebration on the one hand, and the association with death, spirits, and the uncanny in the context of Halloween, on the other.

To take advantage of the particular qualities of moonlight, some gardeners plant so-called moon gardens, with evening or night-blooming plants such as the intensely fragrant and twining moonflower, the shrublike four-o'clock or "beauty of the night," and the toxic angel's trumpet. In wintertime, "Harry Lauder's walking stick," a bizarrely twisted white-stemmed shrub with a mysterious air, is also suitable for this kind of

garden. Viewed by the light of the moon, such a collection of plants exudes a special dreamlike atmosphere. The moonlight enhances the sight of the blooms, which stay open only until the morning sun touches them. This odd pattern of display protects the flowers from the heat of the day and allows them to attract by their scent nocturnal insects for pollination. Red roses would make little contribution to such a garden: moonlight would simply turn them gray.

Geisha strolling by moonlight (1887), an illustration from Tsukioka Yoshitoshi's (1839–1892) *One Hundred Aspects of the Moon*

CHAPTER FIVE

Encounters of a Lunar Kind

We have seen the speculation by scientists about the possibility of life on the moon, and their later confrontation of the reality. Writers of fiction took many more liberties with the facts, but their imaginative moon cultures were not formulated arbitrarily. These literary fantasies are products of particular times and circumstances, sometimes incorporating newly emerging technical insights and possibilities, sometimes reflecting philosophical thought about what a perfect world should look like or how a less perfect world might be construed: the fictional moon became utopia or dystopia, or something in between. Or, as Scott L. Montgomery has expressed it, the moon "served as a historical sponge for sensibilities then percolating through European society."

Speculation about life on the moon long precedes the telescope, of course, going back at least to the philosopher Philolaus (ca. 470–ca. 385 B.C.), who believed that the moon had mountains, valleys, humans, animals, and plants that were much taller and more beautiful than those on Earth. Because of the longer duration of the lunar day, Philolaus assumed that the humans there would be fifteen times as large as those on Earth. Plutarch, for his part, not only believed

This is how Gustave Doré imagined Lucian of Samosata's tale about a trip to the moon.

the moon to be of an earthen substance but also thought there were people on the moon and that the souls of the deceased go there.

In 1686 the French mathematician Bernard de Fontenelle (1657-1757) published his *Discourses on the Plurality of Worlds.* Apart from popular explanations of the heliocentric model of the universe, this work contained many speculations about life on other planets in the solar system. Fontenelle thought the lunar air too rarefied to support life, but he nonetheless portrayed the moon as a sort of utopian alter ego to the Earth. For Fontenelle the moon was only the first of a series of second worlds the newly discovered universe must contain. Furthermore, the developing new model of the cosmos, which abandoned the medieval concept of crystalline spheres separated from one another, made space travel a theoretical possibility after all.

How did philosophers and writers imagine these first explorations of the moon? How did they envision the Selenites or Lunarians—as the moon's inhabitants were called—and their world? Would moon dwellers resemble humans? In *True History,* the Greek Lucian of Samosata (ca. A.D. 120-180) describes a whirlwind that lifts a sailing vessel from the ocean and carries it to the moon. He calls the lunar inhabitants Hippogypi and portrays them as riding on three-headed vultures with unusually large feathers. In another work with a similar theme, Lucian's hero Icaromenippos reaches the moon merely by flapping his arms after attaching the wings of a vulture and an eagle, "to make a bird of myself, and fly up to heaven." The feat is possible, in Lucian's scheme, because he assumes an unbroken atmosphere of air, based on Aristotle's dictum that nature abhors a vacuum—an error that colored scientific thought for almost two millennia. Icaromenippos takes off from Mount Olympus and, after three days, comes to rest on the moon. From that vantage point he can observe the crimes on Earth, and he meditates on the great universe in front of him.

The next known imaginary trip to the moon is centuries later, in Persian literature. The epic poem *Shāhnāmeh,* written by Firdausi and published in 1010, incorporates orally transmitted legends into an account of a marvelous flight to the moon and through the heavens. Five centuries later, the Italian Ludovico Ariosto (1474–1533), in *Orlando Furioso* (1516), an epic poem in four cantos, told of four steeds "redder than flames" that transport the British duke Astolpho, accompanied by John the Evangelist, to the moon. It is both a place for contemplation and a fantastic setting, a stopover for events past and still to come. "Swell'd like the Earth, and seem'd an Earth in size," it had cities and castles just like our planet.

The eminent German astronomer Johannes Kepler (1571–1630) knew that there was no atmosphere between the Earth and the moon, so the use of animals for the trip was out of the question. He could not provide a plausible means of travel to the moon without bringing supernatural forces into play. In his tale *Somnium* (The dream, 1634) a demon—abhorring sunlight, but able to travel at night—makes possible the four-hour trip from Volva, the Earth, to the island Levania, the moon. The beings envisioned by Kepler lived in caves and crevices, emerging only briefly during the day.

The Man in the Moone; or, A Discourse of a Voyage thither, written by Bishop Francis Godwin of Hereford (1566–1633), is the earliest surviving moon travel story in the English language. Godwin's hero, a certain Domingo Gonsales, is a good Spaniard of a poor family. Marooned on an island, Saint Helena, he tames a kind of wild swan called a gansa, which has one webbed foot and another with eaglelike talons. Gonsales constructs a contraption with a seat to be carried by the tamed birds. Unaware that the gansa is migratory, and accustomed to wintering on the moon, he becomes an unwitting passenger on the birds' annual migration. As they ascend, he feels the Earth's gravity diminishing. He is surprised to find his birds flying "as easily and quietly as a fish in the middle of the water . . . whether it were upward, downward or sidelong, all was one." En route to a safe landing atop a lunar hill, Gonsales observes that the Earth rotates around its own axis—just as he had been taught as a young student in Salamanca, the first university to introduce the heliocentric cosmology of the Polish astronomer Nicolaus Copernicus (1473–1543). But Gonsales doesn't go as far as the famous astronomer and declare that the sun is the center of the universe.

Francis Godwin posits his "second Earth" as a utopia whose natural environment makes "it seemeth to be another Paradise." Godwin portrays a surface covered by a huge sea with trees three times as high as on Earth and more than five times as thick. The moon is inhabited by a race unlike any an earthling has ever encountered. Their skin is "lunar" in color, "having no affinitie with any other that ever I beheld with mine eyes." Happy and satisfied, these people display no feelings of hatred or envy; fighting and murder are unknown to them. They cheerfully take Gonsales by the hand and bring him to their homes. He crosses himself and, amazed by this encounter, invokes the names of Jesus and Mary. The immediate reaction of his hosts fills him with delight: "No sooner was the word Iesus out of my mouth, but young and old, fell all downe upon their knees (at which I not a little rejoyced) holding up their hands on high, and repeating all certain words which I understood not." The Lunars, as he calls them, sleep whenever the sun is shining, and none live on the far side of the moon. Observing their fondness for tobacco, he assumes a connection to the Native Americans; in fact, he believes the Lunars are their descendants.

The opening sentence of Bishop Francis Godwin's *The Man in the Moone*

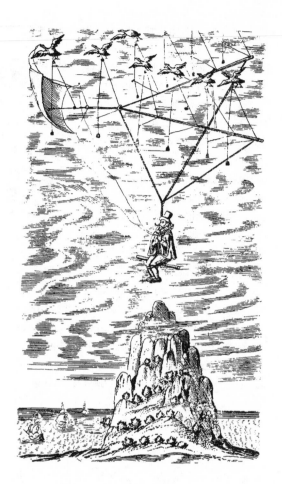

Gonsales's flight to the moon with the help of gansas

On Earth, Gonsales had to learn Italian, French, and German to make himself understood. The Lunars, though, speak the universal language from before the construction of the legendary tower of Babel. Not only is this language mellow and melodic, it is entirely made up of music and therefore superior to any spoken language. The Lunars, quite simply, have "music in themselves."

This aspect of the tale reflected the search among seventeenth-century scientists for a universal language. Some believed it had been located in China, where Jesuit missionaries reported the use of signs and symbols as a kind of picture language intelligible all over the country, despite the various spoken dialects. And indeed, according to Gonsales, the language of the Chinese is comparable to the one spoken by the inhabitants of the moon: the *lingua humana,* the beautiful language of Adam. Soon the traveler becomes homesick, and the gansas are due for their return migration. Gonsales harnesses the birds for the return to Earth and, after nine days, crash lands in China, narrowly escaping execution as a sorcerer.

Lunar inhabitants in the imaginative accounts are hardly ever imagined as inferior, ill-natured, or threatening; rather, they often possessed superior qualities rare on Earth. Reports of encounters with "primitive" people on real voyages of discovery on our own planet certainly helped inspire these fictional accounts.

The French satirist Cyrano de Bergerac (1619–1655), in *The Other World: The Comical History of the States and Empires of the Moon* (1657), suggests equipping the roof of a carriage with a firework in order to reach the moon. De Bergerac's moon is "a World like ours, to which this of ours serves likewise for a Moon." But the satiric portrayal calls into question the foundations of society. Old people obey the young, birds talk rather than sing, trees philosophize, and payment is made with self-written poetry rather than coins.

An illustration for Cyrano de Bergerac's tale

The noble lunarians live on the vapors of cooked food rather than food itself and communicate with simple melodies that combine to produce harmonious concerts.

Imagination had no bounds. In *Iter Lunare; or, Voyage to the Moon* (1703) David Russen employs just enough science to acknowledge the thin atmospheric layer of the Earth and the decrease of weight with distance. He dismisses all the vehicles proposed for trips to the moon—including de Bergerac's rocket carriage—and introduces a spring catapult to hurl the visitors to the moon and back again.

In Murtagh McDermot's *A Trip to the Moon* (1728), the protagonist is gripped by a whirlwind at the Pico de Tenerife on the Canary Islands and soon reaches "a Space between the Vortices of the Earth and moon, where the Attraction of neither prevail'd, but the contrary Motions of their Effluvia destroy'd one another." He simply clings to "a Cloud full of Hail" and moves "with incredible swiftness" into the sphere of attraction of the moon until he falls toward it, luckily into a fishpond. After many experiences there, he prepares himself for the return to Earth: "I design to place myself in the middle of ten wooden Vessels, placed one within another, with the Outermost strongly hooped with Iron, to prevent its breaking. This will be placed over 7000 Barrels of Powder, which I know will raise me to the top of the

Atmosphere." Before digging a hole in the moon's surface for the gunpowder to be "blown up," he fills the spaces between the vessels with water to prevent a fire. Once he is aloft, woodcocks encountered en route guide him home.

In *The flying tank* (1783), by the pseudonymous Madam la Baronne de V***, the voices of the moon's inhabitants have an angelic beauty reminiscent of flutes. These pure, sweet-smelling beings take nourishment from the water of a river. They immediately subject their malodorous guests from the Earth to an intensive cleaning. The visitors are then led through a complex cave system, where they encounter allegorical figures representing traits, positive and negative: false love is manifest as envy, jealousy, seduction, and deceit; real love finds itself side by side with confidence, security, and sweetness.

In the anonymously published *Interesting account of a new trip to the Moon* (1784), this "so much desired world" is a peaceful place without any wild animals. A single variety of fruit provides all food and drink, and table manners require a diner to laugh wholeheartedly when the host is drinking. Male inhabitants are ugly, but the females are beautiful, and they create an erotic ambience by exposing their breasts to the visitors. While the lunar women are permitted a certain degree of tenderness with their guests, the familiarity is limited to a kiss; they always stay faithful to their spouses.

Such lunar utopias were hardly limited to France or England; they can also be found in Russian literature during this time. In Vasilii Levshin's *The Latest Journey* (1784) the moon is a world of absolute equality with neither soldiers nor sovereigns, where tradition rather than progress rules and the inhabitants devote themselves to such romantic activities as farming and sheepherding. Ironically, for Levshin the "lunatics," the people in the moon, are the only sane people of the universe. In a similar vein, Mikhail Chulkov in *Dream of Kidal* (1789) portrays a moon world in which all property is held in common and creatures such as vipers, crocodiles, and tigers live in complete harmony with man. In fact, the moon is posited as being as different from the Earth as paradise from hell. While these were novels, not polemics, their utopian visions certainly reflected the authors' dissatisfaction with contemporary ideals.

The protagonist of George Fowler's *A Flight to the Moon; or, The Vision of Randalthus* (1813) casts up his eyes to behold "a milk white cloud." Examining it further, he realizes that it shrouds a beautiful, angelic woman whose "complexion was as white as the soft descending snow; her cheeks and lips were tipped with pink, her eyes were as bright as the sparkling diamonds." She then opens her sweet lips, saying, "Thou shalt see with thine own eyes that object which has so often been the subject of thy solitary meditations, and behold thou art destined to visit the Moon!" The woman then disappears in a vivid flash of light, but our hero finds himself within the very cloud in which she had originally appeared, "swiftly advancing above the confines of the Earth." The human inhabitants of the moon bear a considerable resemblance to this captivating messenger: "Their complexion is of a beautiful golden cast; their cheeks and lips are tipped with a lively red; their eyes blue; and their golden hair oft falls down their shoulders in beautiful ringlets." Fowler's protagonist observes "great symmetry and delicacy in their shape and features" and notes, "they move with inimitable grace." He remarks on their sensibilities, "They are feelingly alive to all the virtues which we possess; but angry looks never distort the beauty of their features, nor even passions, pollute the purity of their hearts. They are extremely quick in their motions, and equally quick of apprehension. They are fonder of music, painting and poetry, than philosophy and abstract studies, which they say only tend to bewilder the mind without either amusing the fancy or adding to the comforts of life."

In Jacques Boucher de Perthes's tale *Mazular* (1832), the main character falls upon a cloud that then transports him to the moon. He begins to spin so fast that he has trouble breathing and has the impression of soaring upward for fifteen days and nights until he can discern "something round and shiny"—a fairytale moon world where he is thrown down with a terrible shock. The lunar humans he encounters have only one leg, one arm, one eye, and one ear each, and no nose at all; everything on the moon is strangely cut in half. He returns home by chance: leaning out a bit too far, he simply falls back to Earth.

Some tales of lunar exploration reflect techno-logical progress. Romantic moonflight and dream-like devices disappear, and imaginary journeys lay more explicit claim to authenticity. Isaac Newton had found that "to every action there is always opposed an equal reaction: or, the mutual actions of two bodies upon each other are always equal, and directed to contrary parts." The jet force, the fundamental physical principle involved in rocket propulsion, is an extension of Newton's law, and Newton himself foresaw that humans of future centuries might fly to the stars under the power of the principle he described. Early science fiction, still centuries shy of the knowledge needed to make space travel a fact, nonetheless began to ex-ploit Newtonian principles of gravity. *A Voyage to the Moon* (1827), one of the first American novels about an interplanetary voyage, was written by "Joseph Atterly"—a pseudonym for the lawyer George Tucker, who was one of Edgar Allan Poe's university instructors. A metallic substance called lunarium is discovered, which, "when separated and purified, has as great a tendency to fly off from the Earth, as a piece of gold or lead has to approach it"; moreover, it is "in the same degree attracted towards the Moon." A mechanism is constructed "to penetrate the aerial void":

> The machine in which we proposed to em-bark, was a copper vessel, that would have been an exact cube of six feet, if the cor-ners and edges had not been rounded off. It had an opening large enough to receive

our bodies, which was closed by double sliding panels, with quilted cloth between them. When these were properly adjusted, the machine was perfectly air-tight, and strong enough, by means of iron bars running alternately inside and out, to resist the pressure of the atmosphere, when the machine should be exhausted of its air, as we took the precaution to prove by the aid of an air-pump. On the top of the copper chest and on the outside, we had as much of the lunar metal (which I shall henceforth call "lunarium") as we found, by calculation and experiment, would overcome the weight of the machine, as well as its contents, and take us to the Moon on the third day. As the air which the machine contained, would not be sufficient for our respiration more than about six hours, and the chief part of the space we were to pass through was a mere void, we provided ourselves with a sufficient supply, by condensing it in a small globular vessel, made partly of iron and partly of lunarium, to take off its weight. . . . A small circular window, made of a single piece of thick clear glass, was neatly fitted on each of the six sides. Several pieces of lead were securely fastened to screws which passed through the bottom of the machine; as well as a thick plank. The screws were so contrived, that by turning them in one direction, the pieces of lead attached to them were immediately disengaged from the hooks with which they were connected. The pieces of lunarium were fastened in like manner to screws, which passed through the top of the machine; so that by turning them in one direction, those metallic pieces would fly into the air with the velocity of a rocket.

The two protagonists then undertake "a voyage, of which the history of mankind affords no example."

Tucker describes life on the "dark hemisphere":

The Sun pursues the same path in the corresponding latitudes of both hemispheres; but being without any Moon, they have a dull and dreary night, though the light from the stars is much greater than with us. The science of astronomy is much cultivated by the inhabitants of the dark hemisphere, and is indebted to them for its most important discoveries, and its present high state of improvement. . . . If there is much rivalship among the natives of the same hemisphere, who differ in the length of their shadows, they all unite in hatred and contempt for the inhabitants of the opposite side. Those who have the benefit of a Moon, that is, who are turned towards the Earth, are lively, indolent, and changeable as the face of the luminary on which they pride themselves; while those on the other side are more grave, sedate, and industrious. The first are called the Hilliboos, and the last the Moriboos—or bright nights, and dark nights.

In the novel *Trip to the Moon* (1865) by Alexandre Cathelineau, some plants are placed inside an airtight *Terrinsule* test spaceship 50 feet tall and with a capacity of 530,000 cubic feet to accommodate the necessary oxygen for the two travelers. The test craft is the basis for the later *Micromégas* spaceship, constructed with the help of the best carpenters, masons, and gardeners, aided by freed slaves. The inhabitants of the moon are attractive, amicable, and emotionally serene; they speak a single sonorous, musical language. There are no murders, wars, or sickness, and therefore no need for lawyers—"a paradise superior to the one of Adam and Eve before the fall." The lunar people build bridges and houses of wood and travel in carriages drawn by mooselike animals or through the air on the back of eaglelike birds. Although they have many common duties, they clearly reject socialism. In a religion reminiscent of a nature cult, they render homage to God with ceremonial fires. The visitors are essentially reincarnated on the moon: they remember no details from their life on Earth.

In "The Unparalleled Adventure of One Hans Pfaall" (1835), Edgar Allan Poe sends his hero to the moon in a hot-air balloon, which was the only real-life means of aerial transport available until early in the twentieth century. As the title suggests, the story revolves around a certain

Hans Pfaal's balloon trip to the moon

Hans Pfaall, who, having killed some of his creditors, goes to the moon on the first of April to escape the rest. After spending five years there, he sends one of the lunar inhabitants back to Earth in his balloon to deliver the message that Pfaall will return if the citizens of Rotterdam grant him full pardon for his crimes. The lunar inhabitant, frightened upon his arrival by the "savage appearance of the burghers of Rotterdam," simply throws Pfaall's note to the crowd and returns home without waiting for a reply. The public regards Pfaall's story as a hoax; the protagonist, for his part, "cannot conceive upon what data they have founded such an accusation."

In August 1835 the *New York Sun* reported in a series of articles under the headline "Great Astronomical Discoveries" that Sir John Herschel had discovered blue bat people on the moon with the help of a very powerful telescope. Herschel had allegedly observed "sheep, pygmy zebra, [and] unicorn roaming lunar grasslands," in addition to the "bipedal winged creatures called manbats." These creatures averaged "four feet in height,"

> were covered, except on the face, with short and glossy copper-colored hair, and had wings composed of a thin membrane, without hair, lying snugly upon their backs, from the top of their shoulders to the calves of their legs. The face, which was of a yellowish flesh color, was a slight improvement upon that of the large orangutan, being more open and intelligent in its expression, and having a much greater expansion of forehead. The mouth, however, was very prominent, though somewhat relieved by a thick beard upon the lower jaw, and by lips far more human than those of any species of simian genus.

These descriptions originated not with Herschel but from the pen of an imaginative *Sun* reporter. Sales skyrocketed, and other newspapers around the world reprinted the stories. The moon hoax

ARTICOLO

ESTRATTO DALLA GAZZETTA DI FRANCIA

SULLE

SCOPERTE FATTE NELLA LUNA

DA HERSCHEL

A specimen of the curious race of bat people conceived by Richard A. Locke

was a landmark in the emerging sensational press of the 1830s and also a kind of litmus test. The strong reaction of the public proved that such ideas still could not simply be dismissed as nonsense; science fiction presented as fact could draw a large, fascinated audience. The publisher of the newspaper boasted, "New Yorkers read the *Sun* by day [and study] the Moon by night." Richard Adams Locke (1800–1871) soon confessed that he had fabricated the story.

For some writers, the fact that the moon is much smaller than the Earth suggested that its inhabitants would likewise be smaller than humans. In *Trip to the Moon* (1845), Jacques Bujault follows this logic: the diminutive size of his lunar *Picolins* is the only characteristic that distinguishes them from humans. In the anonymous *Very Recent Trip to the Moon* from the same year, the vegetarian inhabitants of this "other eden" reigned over by God himself are depicted as attractive beings with blond hair and blue eyes,

"men, women and children smaller than us, but of a most charming physiognomy, full of expression, of grace, of happiness." Born innocent, they await a different fate from ours, happily celebrating the arrival of an angel to take away a deceased body, as decay is unknown there.

Some authors presented a moon that would be familiar to any earthling. Georges Le Faure and Henry de Graffigny, in *Extraordinary Adventures of a Russian Savant* (1889), imagined a lunar far side with trees and forests where the

The most beautiful of the selenites in Georges Le Faure's and Henry de Graffigny's novel *Extraordinary Adventures of a Russian Savant* (1889)

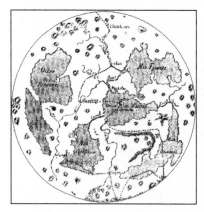

The map of the far side as provided by Le Faure and de Graffigny

giant Selenites were concentrated. The authors even provided the readers with a map of the far side, showing not only the elements typical of the visible side of the moon, such as craters, but also oceans, rivers, and cities. Other writers twisted familiar values to create their lunar fables. In Louis Desnoyer's *The Adventures of Robert Robert* (1839) the moon is an Earth where everything is topsy-turvy. Tiny elephants populate the cracks and crevices instead of ants, and sheep keep flocks of wolves together. Although gold, silver, and diamonds don't have any value, iron and pebbles are expensive. And since the rain is made of wine, water is the beverage most in demand.

The pseudonymous Pierre de Sélènes creates a vaguely communistic utopian society with a naive belief in technical progress typical of the time in his novel *An Unknown World: Two Years on the Moon* (1886), dedicated to Jules Verne. Under the surface of the moon three travelers from Earth discover a naturally created, weatherless, "supernatural" world inhabited by some twelve million people. The world has an ocean the size of the Mediterranean at its center and a delightfully beautiful city with richly ornamented houses, illuminated by cosmic and electric light. A variety of stone functions as a battery, taking its charge from the electricity-laden air. The lunar people—

who take nourishment from the air, never kill, and speak a musical language "of an extreme logical simplicity"—had withdrawn into this underground haven when the water and air on the surface began to disappear. Wages and private property have no place in this ideal society, and every citizen can claim the space he needs. Futuristic trains provide transportation through the sublunar cave world, and the inhabitants also use technology that suggests television: "transmis-

A lunar base from the book by Pierre de Sélènes

sion at distance of sensible and talking images,"
technology still decades away on real-life Earth.
Outside, they are able to witness volcanic explosions that spew into the sky large incandescent chunks of matter, which, once they have
returned to the thin and cold lunar atmosphere,
simply turn into flying sparks—a natural catastrophe "nothing on Earth could have given the
slightest idea of." But the visitors finally decide
to return to Earth, partly because they are bored
by the day-to-day monotony and the unchanging weather, partly because they find the sublime
character of the lunarians too distressing.

Fantasies about lunar life certainly entertained
readers of earlier centuries, but they also func-
tioned as blueprints for speculation about life on
the planets. Of course, the fantasies evolved along
with the ongoing technological progress. As evidence mounted that life on the moon was unlikely, if not impossible, speculation shifted elsewhere—toward Mars, for example, where similar
fantasies took root. But lunarians survived in a
few exceptional works of twentieth century science fiction—both novels and movies. As we'll
see in a later chapter, this generation of moon-
men is not quite as sweet and lovable as earlier
authors had envisioned.

In André
Laurie's science-
fiction novel
*The Conquest
of the Moon*
(1875) the
moon is drawn
from its orbit
to land in the
Sahara desert.
Laurie was the
pseudonym of
Jean-François
Pascal Grousset
(1844–1909).

CHAPTER SIX

Lunar Passion in Paris

The European fascination with the moon in the eighteenth and nineteenth centuries was part of a deep obsession with wild and sublime landscapes. Jagged volcanic vistas, the icy regions of the Arctic and Antarctic, and the dry vastness of the deserts attracted scientists and laymen alike, some of whom paid for their curiosity with their lives. All such environments imposed exacting physical limitations on the humans willing to investigate them. As Alain de Botton has put it in his essay *The Art of Travel,* "It is as if these landscapes allowed travellers to experience transcendent feelings that they no longer felt in cities and the cultivated countryside." The key difference from the other landscapes was that the moonscape was inaccessible to the whole spectrum of the senses, could be explored only visually, and still remained largely a landscape of the imagination.

The French writer Jules Verne (1828–1905) was gripped by this fascination. Though not a scientist himself, he had read popular works of astronomy, had a large archive, and kept current with scientific and technological developments. Until the first balloon flight, in 1783, most people saw little difference between flying above the surface of the Earth and traveling to other celes-

Jules Verne's spacecraft on its way to the moon

Jules Verne

tial bodies—both seemed utterly impossible. In contrast to many earlier tales of trips to the moon, Verne's *From the Earth to the Moon* (1865; English translation, 1867) seemed more realistic, because it took contemporary scientific knowledge into account—particularly ballistics, the scientific basis for understanding projectiles in flight, which had made many advances. The novel also reflects contemporary scientific disputes about the possibility of volcanic eruptions on the moon or whether the moon's hidden side might have air, water, clouds, plants, or even a forest. Since the development of mechanically powered aircraft and the drama of a real trip to the moon were a long way off, Verne's account was still an exercise in imagination. But the idea no longer seemed as implausible as it had a century, or even a couple of decades, before.

Like other books Verne wrote, *From the Earth to the Moon* is startlingly accurate in anticipating technological developments of the coming century. (Verne even had an agreement with his publisher to incorporate the most up-to-date scientific facts.) The book is unprecedented in its systematic treatment of scientific elements of preparations for the trip. It tells the story of three members of the Baltimore Gun Club who plan to build an aluminum cannon nine hundred feet long and six feet wide and launch themselves to the moon using gun cotton. Verne thus introduced the idea of a manned ballistic projectile escaping the Earth's gravity.

Although from our perspective Verne failed to consider some technical issues, such as the friction on the projectile during flight, he was the first to identify other salient aspects of space flight. Some details of the story even bear similarities to the National Aeronautics and Space Administration (NASA) Apollo program of the 1960s, beginning with the number of astronauts, three. The second volume of the work was serialized in 1869, exactly one hundred years before the first landing of humans on the moon. Verne correctly predicted not only that the United States would launch the first manned vehicle to orbit the moon—and would strive to be the first country to plant its flag there—but also that Florida would be the launch site. He selected Tampa, near the Gulf Coast, whereas NASA chose Merritt Island, 130 miles away on the Atlantic side. Both sites have easy access to the sea and fulfill the criterion of being at a latitude below 28 degrees north. Modern space projects favor sites relatively close to the equator because the rotational speed of the Earth is greatest there, providing a small extra boost at launch. But Verne chose this site for a different reason: to facilitate coordination of the launch with the moon's perigee—the point at which it is closest to the Earth. Verne can thus be credited with anticipating the idea of a temporal and spatial frame for the launch of a rocket or spaceship. In addition, he reasoned that a large body of water would be the safest place to land. Both Verne's imaginary moon mission and the Apollo flights splashed down in the Pacific Ocean and were rescued by U.S. Navy vessels. In both the novel and in reality the start of the rocket drew an immense number of people to the site. Before the invention of the automobile, Verne still expected onlookers to arrive by railroad.

Verne imagined a spacecraft constructed primarily of aluminum and weighing 19,250 pounds, whereas the predominately aluminum *Apollo 8* circumlunar spacecraft weighed 26,275 pounds when empty. The cannon used to launch Verne's spacecraft was called *Columbiad;* the command module of *Apollo 11* was the *Columbia.* The invocation of Christopher Columbus likens the trip to the moon to the explorer's ambition to find a new route to the Indies, which resulted in the discovery of the New World. That connection touches on a psychological similarity between Verne's travelers and American astronauts: both groups understood themselves as representatives of Earth, and both were preoccupied with humanitarian and even religious concerns. The journey of Verne's spacecraft took 242 hours, 31 minutes, whereas the *Apollo 8* crew spent 147 hours in space. Considering the enormous leaps in knowledge that occurred in the hundred years between the fictional and real space flights, many of Verne's predictions are stunningly accurate.

In some areas, Verne was on less secure scientific footing. It is vital, for example, to consider the effect of gravitational force on humans aboard a giant projectile accelerated instantly to a speed capable of escaping the atmosphere. Verne

was aware of this problem, but the hydraulic shock absorbers he provides for the hollow shell would not have had the desired effect. Instead, the tremendous pressure would have crushed everything—including the passengers—like a soap bubble. Verne also erroneously believed that weightlessness would occur only for a short time at the neutral point of gravity between the Earth and the moon.

In the novel a meteor sidetracks the projectile from its planned course so that instead of falling

Interior of the cannon designed to shoot a spacecraft to the moon in *From the Earth to the Moon* (1865)

on the moon, it traces an ellipse around it. Missing the moon is actually a stroke of luck for the imaginary travelers—given the logistics of their space travel, had they landed, they would not have had the slightest chance of getting back. By not actually landing on the moon, they also spared Verne the need to describe a lunar environment about which little scientific consensus existed. He didn't have to stick his neck out too far on this point. While circling the moon, his space travelers believe they recognize ruins on the far side, even canals, but they find no clear evidence of lunar life. Had his travelers reached their goal, Verne would have been forced to pin himself down to one theory. Rather than take the risk that future developments would prove him wrong, Verne simply had his characters repeat the various competing hypotheses about the moon's nature.

We may also see the "test flight" that the adventurers conducted with a cat and a squirrel as a curious precursor to later actual flights using dogs and chimps to study the effects of weightlessness. But Verne errs when he assumes that the travelers could dispose of the body of Satellite, a dog accidentally killed during the trip, "just as sailors get rid of a dead body by throwing it into the sea." In Verne's account, the travelers simply open a hatch in the capsule "with the utmost care and dispatch, so as to lose as little as possible of the internal air," and launch the poor dog's body from their luxuriously cushioned projectile into space. In fact, such a breach of the capsule's atmosphere would have killed all the inhabitants almost instantly.

Were all of Verne's accurate or nearly accurate predictions just strange coincidences, or can we credit him with an anticipatory genius, a special sense for the most likely course of scientific development? Certainly, the era during which he wrote his novel was conducive to innovation in theory and practice, and Verne's fictional account captured the popular imagination. Even though a trip to the moon long remained unlikely in real life, the possibility, however slim, had taken a firm and permanent hold on the cultural imagination. Verne's seminal work inspired most of the original pioneers in the fields of astronautics and space sciences. His novels were widely viewed condescendingly by his contemporaries, many of whom considered them unfit for adult readers; poor translations hampered the reception of his books in other countries until the mid-twentieth century. The Apollo program not only helped make Verne popular again but also led to a new appreciation of the scientific and literary merits of his work.

Although for decades Verne's work was subject to a curious combination of popular success and literary snub, he now has been accorded an undisputed place among the writers in the history of science fiction. Many of those who later made space travel a reality are known to have devoured Verne's novel, so that perhaps more than any other work of science fiction—to an extent that even a bright and imaginative mind such as Verne's would not have been able to anticipate— *From the Earth to the Moon* has had a revolutionary impact on real-world science.

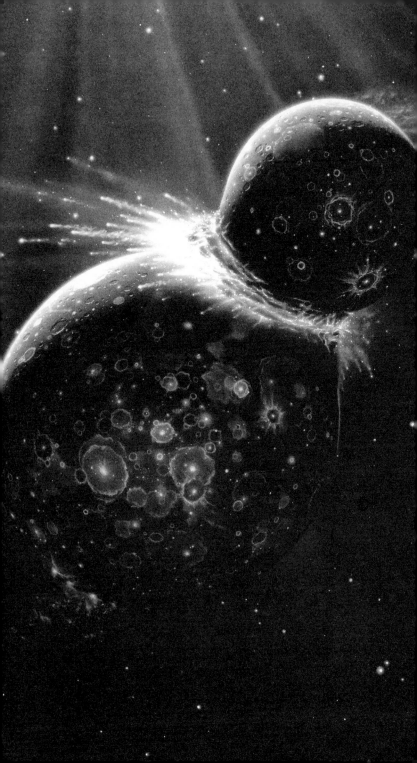

CHAPTER SEVEN

Accounts of Genesis

What is the origin of the moon? A simple question, perhaps, but one without a simple answer. A myth among the Native American Seneca tribes claims that a wolf sang the moon into existence. In the Western world, the study of the history of the Earth was, until the early eighteenth century, still dominated by the biblical account of the Creation, and the same account settled the history of the moon. According to Genesis, God "set two great lights in the firmament of heaven, the greater light to rule the day and the lesser light to rule the night." Just as the moon took its light from the sun, and was inferior to the sun, so secular authority, whether royal or scientific, reflected religious, thus papal, authority. But the European Enlightenment signaled that God's power was beginning to wane.

By the seventeenth century a revolution in thought was afoot, a revolution with reason, not divinity, at its center. That upheaval brought with it a new understanding of the cosmos. The French savant René Descartes (1596–1650) not only put the sun in the middle of the universe, as Copernicus and Galileo had done before him, but also imagined a universe with pieces of matter of different sizes, forms, and motions caused by different forces. This concept put the moon and its

The big splat—one theory of moon formation—as illustrated by the space artist Joe Tucciarone formation in a new framework. In fact, Descartes explained, the vortex of the Earth traps the moon and thus forces it to move around it. Descartes wrote this treatise about 1630, but its publication had to wait until 1664, well after his death, precisely because it affronted the older order. We should not forget that in 1633 Galileo's heliocentric system was condemned in a trial, and he was put under house arrest for the remainder of his life.

Toward the end of the eighteenth century, Pierre-Simon Laplace (1749–1827) proposed that the solar system had begun as a huge swirling cloud (or nebula) of gas. As it collapsed under its self-gravitation, the central mass became the sun, while rings ("Laplacian rings") of material were spun off in succession, from the outer regions first; this matter then coalesced to form the planets. Earth and moon had remained connected in terms of gravitation ever since. This model became the basis of the scientific theory referred to as co-accretion or binary accretion and was also connected to hypothesized changes in the lunar orbit—for example, that the moon is gradually receding from the Earth and must thus have been closer in the past.

George H. Darwin (1845–1912), the second son of Charles Darwin and in his time one of the world's eminent physicists, calculated that at some point the center of the moon would have been only six thousand miles from the Earth's surface. Darwin thus concluded "that if the Moon and Earth were ever molten viscous masses, then they once formed parts of a common mass." He then proposed in 1878 that the Earth had spun so rapidly in its early years that it grew increasingly elongated under the influence of the sun's tidal force until a chunk eventually broke free. This chunk, Darwin argued, became the moon. His grandfather Erasmus Darwin (1731–1802) had already posited the ejection of matter from the Earth to become the moon in a lengthy poem, *The Botanic Garden* (1792):

> Gnomes! how you shriek'd! when through
>   the troubled air
> Roar'd the fierce din of elemental war;
> When rose the continents, and Sunk the
>   main,
> And Earth's huge sphere exploding burst
>   in twain.
> Gnomes! how you gazed! when from her
>   wounded side
> Where now the South-Sea heaves its
>   waste of tide,
> Rose on swift wheels the Moon's refulgent
>   car,
> Circling the solar orb, a sister star,
> Dimpled with vales, with shining hills
>   emboss'd,
> And roll'd round Earth her airless realms
>   of frost.

Erasmus Darwin was not unique in choosing verse as the forum in which to express scientific thought, but his theory was not circulated widely among scientists, even though he was a member of the Lunar Society. And his mention of the South Sea notwithstanding the English geologist Osmond Fisher (1817–1914) is usually credited (or blamed) for the sensational addendum to this fantastic theory that the Pacific Ocean basin marks the "scar" left behind where our future satellite was ripped away. According to Fisher, the American continent was also separated from Europe and Africa during this process. The American geologist and chemist Richard Owen had already suggested in 1857 that the moon split off from the Mediterranean. Such ideas, though later proven to be wrong, foreshadowed the research in the early twentieth century by Alfred Wegener (1870–1930) on plate tectonics and continental drift. George Darwin didn't adopt any of these addenda, but they nonetheless became part of popular accounts of the moon's origin.

William Henry Pickering extended Darwin's lunar hypothesis, speculating that a former single continent consisting of America, Asia, Africa, and Europe had broken apart because of the moon's separation. As the so-called "fission theory" evolved, the assumption arose that the separation had occurred when the Earth was still in fluid state; the matter ejected, according to this line of thought, first formed a ring orbiting the Earth before combining into what became the moon.

George Darwin's theory was eventually replaced by a competing one. This one held that the moon had been formed in a completely foreign realm, possibly even outside our solar system, and was then "captured" by the gravitational pull of the Earth and kept within its orbit ever since. This hypothesis is generally associated with the American chemist Harold C. Urey (1893–1981), who received the Nobel Prize in 1932 and collaborated on the atomic bomb. Soon after World War II he pioneered the field of cosmochemistry. Urey saw the moon as a cold relic from the early history of the solar system. He based his argument on the different densities of sun and moon and the assumed similarity in iron of both bodies. In fact, he drew the conclusion that the moon was never a part of the Earth but had formed independently, and earlier. Urey's notion that the moon had been captured by the Earth provided a perfect scientific rationale for the Apollo program: if the moon predated the Earth, investigation of the lunar surface could produce results that could not be drawn from the geologically active Earth. He also suggested that the moon had become cold since the capture, and that the lunar maria had been formed not by lava flows but by water, possibly splashed from Earth during the moon's capture. "If indeed the surface of the moon carries a residue of the ancient oceans of the earth at about the same time that life was evolving, the Apollo program should bring back fascinating samples which will teach us much in regard to the early history of the solar system, and in particular with regard to the origin of life," Urey remarked in 1966. "I wish I could go rock-hunting with the astronauts this month," Urey said in the summer of 1969, and his addendum, "even if I knew I could never get back," proves the urgency these questions had assumed

Harold Urey

for him. His concrete influence on the space program remained minimal, though.

The three hypotheses of lunar origin available at this decisive time could be simplified in family terms: the moon as Earth's sister, born nearby out of material similar to the Earth's (co-accretion theory); the moon as Earth's daughter, born directly of our planet (fission theory); or the moon as Earth's wife, born elsewhere in the solar system and later captured into the Earth's orbit (capture theory). But by the end of the 1970s, lunar scientists agreed that none of the three theories could offer a satisfactory narrative of the moon's origin. The fission theory had been refuted, because the Earth had not been spinning fast enough to throw off part of itself, but the inescapable paradox was that the chemical compositions of the moon and the Earth held certain similarities. Key elements in the composition of the Earth—oxygen, silicon, aluminum, calcium, magnesium, iron, and sodium, to name just the most important ones— are all there on the moon, albeit in different concentrations. How could this issue be resolved?

Earlier voices had suggested a key theoretical difference, but few had wanted to listen. As early as 1911, the American geologist Howard Bigelow Baker had speculated that the Earth had once had a close encounter with a now-vanished planet, and that what is now the moon had been ripped from the Pacific Basin, thus causing a continental drift. In the 1970s the engineer James G. Baker (no relation to Howard B. Baker) suggested that Mars and Earth had narrowly avoided collision after a massive proto-Jupiter had disturbed their orbits, but his theory was widely ignored.

Before Apollo the moon had been believed to be just a rock—a relic of the solar system's formation that never formed a core. William K. Hartmann and Donald R. Davis hypothesized in the early 1970s the impact of one or more objects six hundred miles in diameter that could have ejected enough material from the Earth's mantle into orbit to form the moon. Hartmann and Davis believed that the colliding planet actually had a molten iron core, but it sank to the center of the Earth and merged with its core, thus leaving the moon relatively devoid of metallic iron. The problem with this theory is that an object six hundred miles across is relatively small—less than 1 percent of the Earth's mass, and less than 30 percent the diameter of the moon itself. Alastair Cameron and William Ward recognized that a much larger object—something with about 10 percent of Earth's mass, an object that was essentially a planet in its own right, roughly the size of Mars today—could have provided the Earth with the rapid initial spin needed to account for our twenty-four-hour day. A major conference in Hawaii in 1984 brought together about one hundred astronomers and resulted in consensus for the giant impact hypothesis.

Alastair Cameron was the first to simulate moon-forming collisions in the 1980s and 1990s. Robin M. Canup from the Southwest Research Institute, employing computer simulations, has since refined the theory with further detail. Her computer model consisted of twenty thousand components, and the collision behavior of each is analyzed. According to her research, the likeliest scenario is that the moon was formed after a glancing collision at high speed occurring four and a half billion years ago between Earth, by then nearly fully formed but smaller than at present, and a proto-planet about the size of Mars. Canup terms the early Earth Gaia and the other body Theia. The collision was the biggest blow Earth ever experienced. The impact would have erased all surface features and probably melted rocks to a depth of 600 miles—"the zero moment" for our planet and the final moment for the other one. Theia, which is believed to have been formed in an orbit similar to the Earth's, was pulverized, its matter sprayed out into a shower of orbiting debris. Within hours much of that debris regrouped to form a new body that smashed into the Earth's surface a second time and was destroyed. Most of the resulting matter was absorbed and became part of our planet, but about one-tenth, mostly from the outer parts of Theia, formed a cloud of debris, an incandescent disk around Earth. This disk was thrown into space and within a few decades developed into what is now the moon. At that distant point in time, the moon was fifteen times as close as it is now. As in Cameron and Ward's scenario, the impact set Earth aspin on its axis, eventually resulting in twenty-four-hour days. On the other hand, the presence of the moon creates a gravitational counterbalance that stabilizes Earth's slightly inclined axis of rotation, providing the congenial cycle of the seasons over a single orbit around the sun. This model also accounts for the moon's comparatively low density and its dry and refractory composition, while also including a source of energy sufficient to melt the upper layers of the early moon and produce an ocean of magma, which, over time, solidified and crystallized into the current hard crust. After the moon formed, about 3.9 billion years ago, a "lunar cataclysm" occurred — a cluster of collisions on the highlands. Rock was driven to the surface by independent eruptions from various sources deep in the hot lunar interior, filling basins with basaltic lava. Some remaining parts of this molten magma stabilized in pockets for some time before erupting to the surface.

Canup's theory includes features of each of the three mutually exclusive classical theories. Accretion plays a role, but the fragments accreted resulted from the collision. As in the double-planet theory, a second planet is involved in the scenario. And, as in the concept of fission, the moon was born of the Earth, but only after a collision with the second planet. The sequence of events as now envisioned was more complex than any of the three earlier theories had suggested.

The giant-impact hypothesis could help clarify why some lunar rocks have a similar composition to those found on Earth and some do not, but it cannot explain why some are magnetized. Perhaps, it has been speculated, the moon has an internally generated magnetic field (as does Earth), or perhaps impact events generate short-lived transient fields that could have magnetized the lunar rocks. Much recent discussion of the moon's makeup focuses on the nature of an assumed core, which may consist of iron with some sulphide. Based on various geophysical calculations, the core is thought to be less than 250 miles in radius—about a quarter of the moon's radius, compared with Earth's core, which is about half of the planet's radius. While it is possible to identify Earth's iron core by means of measurements of seismographs distributed around the planet, analogous equipment on the moon is insufficient to yield such unequivocal information. It is known that moonquakes often occur, but none of their ray paths go through the center of the moon, further complicating core measurement.

Larry Taylor, director of the Planetary Geosciences Institute at the University of Tennessee in Knoxville, has studied mare basalts, which are believed to have been created by melting in the moon's mantle and have retained signatures of that region. The analysis suggests that the lunar mantle has even lower levels than Earth's of elements that bond easily with iron—the so-called siderophile elements, such as platinum, iridium, and osmium. "What happens during the formation of any terrestrial planet is that it undergoes a melting state early in its formation," Taylor says. "In that state, you have separation of metallic iron into a core." When cores formed in Earth and other planets, these elements with an affinity for iron were, for the most part, scavenged from the mantle and transferred into the metallic core. This would explain the relative lack of these elements in both Earth's mantle and the moon's— and in return make the case for a metallic core. The depletion of these siderophile elements in the moon's basalts has been interpreted as additional proof of a metallic core in the moon.

More recently, a team at the MIT led by Ian Garrick-Bethell examined the magnetic history of samples from one of the most ancient rocks in the Apollo collection and found out that the moon in its early days may have had a liquid metal core that spun like a dynamo to produce a magnetic field. The researchers also ruled out that the rock received its magnetic properties from impacts. "This rock was heated up only twice in its history," says Garrick-Bethell. "When rocks cool, they lock in the magnetic fields around them." It remains unclear why such a magnetic field might have died off. In any case the evidence of the lunar core remains far more speculative than scientists like.

One thing is certain, though: moonstones or selenites—gems found in various locations, including Sri Lanka and India, and once believed to interact with the moon—could never have been created there. Moonstones are feldspars, and there are no feldspars on the moon.

CHAPTER EIGHT

A Riddled Surface

We have seen the various theories scientists have come up with to account for the existence of the moon, but what happened after the moon was born? With no time machine available to us, we have to rely on the measurement of radioactive elements in lunar rock samples to determine age. Predictably, the samples differ in age depending on the exact places they were taken from, but virtually none of this material is younger than 3.6 billion years, at which time a heavy bombardment of meteorites ceased. Since then the moon has remained virtually unchanged, though a relatively steady rain of mostly smaller meteorites has continued, over time, to add new soil to the moon's crust.

Aside from the well-known lower gravity relative to Earth, what does it *feel* like to stand on that crust? The lack of a familiar atmosphere means that the sky remains dark even during the long lunar day. In fact, light and dark are turned upside down from our experience on Earth: brownish gray soil contrasts sharply with the blackness of space. The glare reflected by the moon's surface constricts the pupils, weakening vision. While vibrations may be felt, total silence reigns.

The absence of atmospheric haze makes it

Landscape around a lunar pole as imagined by Ludek Pesek (ca. 1960) much more difficult to estimate distances on the moon, as well. We get a feel for how far one spot is from another only when we move, thereby shifting our perspective in parallax displacement. Whatever the actual distance, the moonscape always appears equally sharp. The color of the surface appears to change depending on the angle from which we view it: light brown when we face the sun, but gray in the opposite direction, black when we examine the soil close up. That the moon is much smaller than the Earth means that we can see only half as far into the distance, because the curvature of the surface is greater. The moon doesn't have seasons and climatic zones like those on Earth and Mars; it doesn't even have wind. With the sun creeping slowly across the lunar sky, the peaks of mountain ranges are illuminated several hours earlier than the plains and valleys, and the horizontal areas of the mountaintops are flooded with perpetual light. Temperatures can change rapidly, not only between day and night but also between sunlight and shadow. One of the most dramatic aspects of being on the moon is seeing how Earth, barely moving, dominates the sky. During the Earth's day, we could see cloud formations and watch the planet pass through phases like those we can see for the moon.

Beyond the physical aspects of the moon, we might ponder more philosophical questions about the moon's nature. Is it a *world* even though it is "cold" and lifeless? Before the invention of the telescope, philosophers were often disposed to regard the moon as a world analogous to our own. The Pythagoreans saw the moon as a second Earth. Frank Sherwood Taylor, in *The World of Science* (1937), painted the paradoxical character of the orb: "The Moon is rightly a melancholy sight as well as a beautiful one. It is a sphere of death—a picture of the state of the planet which air and water have left: it revolves, the skeleton of a world." A recurring theme in the literature is that Earth, stripped of water and its mantle of vegetation, might look to a distant observer like the moon.

Given the familiarity of our home planet, it is not surprising that observers have often focused on topographical features of the moon that seem similar to those of Earth while overlooking those that are dissimilar. Circular, craterlike outlines, for example, characterize both the Sea of Japan, between Japan and the Asiatic mainland, and the Gulf of Mexico. The island arcs of the northern Pacific could be compared to the mountain ramparts of the maria. And terrestrial plains ringed by mountains—in Thessaly, Greece, for example—resemble such areas on the moon. But circular features, exceptional on Earth, are more nearly the rule on the moon, whose surface, in fact, is not much like our planet's. The moon's Alps, Caucasus, and Apennines seem at first glance comparable to terrestrial mountain chains, but they probably resulted from faults triggered by mare-forming impacts. Oceanus Procellarum, which measures some 2.3 million square miles, has been called the Pacific of the

A "lunar" landscape: extinct volcanoes of Auvergne, France

moon. That comparison rather shortchanges our Pacific, which is almost thirty times as big.

Similar false analogies have occurred in reverse. Various landscapes on Earth have been characterized as resembling the moonscape: for instance, the Valle de la Luna (Valley of the moon), with its distant ring of volcanoes, in the Chilean region Antofagasta; the volcanic landscape of central Lanzarote, one of the Canary Islands; and the summit of Mont Ventoux in Provence, which has only sparse vegetation and is a major challenge during the Tour de France bicycle race. The Mountains of the Moon, a mountain range straddling the border between the Congo and Uganda, supposedly owe their name to the snow-covered summits somewhat reminiscent of the bone-white lunar surface as seen from Earth. We may also remember the uneven terrain of the Icelandic volcano field of Askja, with its weird rock formations and lava fields, which NASA classified as the most moonlike landscape on Earth, sending Neil Armstrong and his colleagues there for training in 1967. And the nightmare of a "postapocalyptic moonscape" has become synonymous in the collective imagination with parts of the Earth destroyed by nuclear catastrophe.

But how, specifically, is the moon's surface different from Earth's? One key difference lies in Earth's changeability: here seismic activity, weather, and erosion have played and continue to play important roles in altering the surface over time. Craters from meteoric impacts in times long past have eroded and are no longer discernible. More than three-quarters of the surface of the Earth is less than two hundred million years old, and virtually nothing on Earth's surface is the same as it was at the time of its formation. In contrast, according to most recent estimates, 99 percent of the surface of the moon is more than three billion years old—same old, same old, you might say.

The character of the lunar surface had long puzzled scientists, starting with the particular reflective properties and composition of the material of the crust. In 1955 the American geophysicist and NASA consultant Thomas Gold (1920–2004) claimed that the astronauts (and any spacecraft before them) would sink into a yards-deep layer of dust. Gold believed that the lack of lunar wind meant that dust had accumulated on the moon for billions of years, making it virtually impossible for humans to scrutinize the surface. Gold soon corrected himself, positing that the dust would be just a few centimeters thick, and direct measurements later proved that his revised estimate was correct. Despite this vindication, other scientists referred jokingly to "Gold's dust" for many years, and in fact a theory similar to Gold's initial estimate has since been seized upon by some in support of their religious beliefs. Some who believe in literal interpretation of the book of Genesis hold that the depth of accumulated dust on the surface proves that the moon originated about 8000 B.C. If it were really as old as scientists claim, these believers say, continuous bombardment by micrometeorites would have left the layer much thicker. They usually cite an estimate of one hundred feet in controversial studies by R. A. Lyttleton in 1956 and Hans Petterson in 1960, ignoring more recent contradictory findings.

Upon investigation it became apparent that the makeup of the moon's surface was more complex than mere "dust." In 1960 the Dutch scientist Jan van Diggelen noticed what he described as "an irregular spongy character" of the lunar surface and suggested that reindeer lichen (*Cladonia rangiferina*) was the closest terrestrial analogue. Arthur C. Clarke at one point expected it to resemble "stale black bread." The Surveyor expeditions confirmed a granular aggregate, with particles from about one-twenty-fifth to four-tenths of an inch, with some larger rock fragments loosely cemented. The porous matter covering the moon's surface, debris produced by the impact of meteorites, came to be called regolith (literally, "blanket of rock"), with the finer particles usually referred to as lunar soil, even though the organic components of terrestrial soil are absent. This material includes crushed rock, glass beads, and even grains of metallic iron—another difference from Earth's soil, where this mineral would soon rust.

All in all, about 850 pounds of lunar rock was collected during the Apollo missions. Scientists in many countries have analyzed this material, which remains an object of study. Most of this rock is stored under low-humidity conditions at the Lyndon B. Johnson Space Center in Houston, where it has been repeatedly subject to theft from interns. Examination of the samples has revealed that Earth and the moon have identical oxygen isotope compositions, which vary substantially from those of Mars and most of the asteroids, but the moon rocks contain few of the volatile elements characteristic of Earth.

A meteoroid flux of mostly tiny particles constantly strikes the moon, and at times larger chunks of debris hit the surface with a great deal of energy. According to one estimate, a football-sized meteorite hits the moon every day. When a large enough meteorite strikes, a new crater results. Paul D. Spudis has described this ongoing process "as a giant sandblaster, slowly grinding the Moon's crust into dust." In contrast, the Earth's surface is protected by the atmosphere, and particles falling from space usually burn up before impact.

More than a decade after the end of the Apollo

program came a surprise that suddenly made the moon appear a little bit more like Earth. Investigations of lunar samples yielded no evidence of water, but in 1994 water on the moon had suddenly become at least a theoretical possibility. The lunar spacecraft *Clementine,* which orbited the moon for two months, provided maps of the polar regions. Some craters there are so deep that their floors are never lit by the sun; temperatures within are low enough—never above minus 270 degrees Fahrenheit—to harbor frozen water over geologic time. Light coming back from radar pulses sent into the shadowed areas seemed to confirm an icy surface. The researchers who analyzed the *Clementine* data suggested that deposits of ice might cover an area of thirty-five to fifty square miles at the moon's south pole. The possibility of ice sparked immediate interest, because the presence of water would make the establishment of a self-sustaining lunar colony plausible. But there was no clear indication where the ice, if it existed, had originated. According to one theory it didn't vaporize during the collision of the celestial body with the early Earth that produced the moon; alternatively, it may have arrived later via cosmic projectiles. In any case, there could not be much of it. Results of the probe were further put in perspective when, in 1997, the Arecibo Observatory—the world's largest single dish radio telescope, based in Puerto Rico—recorded similar effects in lunar regions at such a distance from the pole that high daytime temperatures make the existence of ice impossible. Apparently, very rough surfaces were producing radar readings similar to those produced by ice surfaces. But Donald B. Campbell, Cornell University professor of astronomy, conceded that since "neither Arecibo nor Clementine observed all the areas that are in permanent shadow, . . . there is still the possibility that there are ice deposits in the bottoms of deep craters." In 1998 scientists working with *Lunar Prospector,* a small, spin-stabilized craft designed to sample the lunar crust and atmosphere for minerals, ice, and certain gases, announced that between ten million and three hundred million tons of water-ice was scattered inside the crater of the moon's poles.

Another surprise occurred in 2008, when Erik Hauri, a geochemist at the Carnegie Institution of Washington's Department of Terrestrial Magnetism, and some of his colleagues discovered that green and orange lunar volcanic glass beads from samples collected during the Apollo missions were rich in some volatile elements and water, which high temperatures during the lunar-formation event may have evaporated. Applying a special spectrometric technique, Hauri demonstrated "that some parts of the lunar mantle may contain a few hundred or even a few thousand parts per million of water." The beads, which are coated with some volatile elements, are thought to have resulted from volcanic eruptions on the lunar surface.

This is not the end of the story, however. This long series of speculation reached a climax in November 2009 with NASA's sensational announcement that a "significant amount" of water had been located in the form of ice that had accumulated over billions of years. A month earlier, a satellite called LCROSS had been intentionally crashed into a crater near the moon's south pole. The impact carved out a hole sixty to one hundred feet wide, releasing at least twenty-five gallons of water to the surface. Does the moon bear yet more mysteries—and greater similarities to Earth—hidden beneath its monochrome surface?

The lunar craters—about 100,000 more than a half-mile in diameter, found mostly on the highlands—have long attracted attention because they present such a contrast with Earth's surface. The depressions can be round, polygonal, or quite irregular, with or without an exterior wall. Craters exist in all sizes: the smallest are submicroscopic, the largest—Aitken Basin at the South Pole—fifteen hundred miles in diameter and seven and a half miles deep. Some have jokingly compared the lunar surface to Swiss cheese, but even serious attempts to classify the various craters that pock the lunar surface have been rather confusing. The very use of the word *crater* is problematic, because the moon's craters are neither of volcanic origin nor as deep as a volcanic crater on Earth. The word's primary utility may be in reminding us of the earlier, outdated, explanation of these formations. The term *impact crater* is now sometimes used to distinguish these from volcanic ones.

Once telescopes became available, scientists quickly began discussing the origin of what

Galileo compared to "eyes upon the peacock's tail." Robert Hooke (1635–1703), secretary of the Royal Society, was one of the first to venture an opinion. *Micrographia* (1665) was devoted to Hooke's discoveries using the microscope, but he digressed to liken the lunar craters to the volcanoes on Earth and thus to attribute them to internal origin. He once dropped musket balls into wet clay, to produce pits resembling lunar craters. At the end of the eighteenth century the astronomer Franz Aepinus of Saint Petersburg also

Earlier selenologists believed the lunar mountains to have sharp pointy edges. This illustration is taken from *Recreations in Astronomy* (1879) by H. D. Warren.

concluded that lunar craters were of volcanic origin, speculating further that basins represent the first stage in the formation of a volcanic mountain. A central hillock of ejected matter forming around the vent he called molfetta. In describing the lunar surface, Aepinus wrote one of the most extreme statements about lunar volcanism:

> Anyone who, armed with this understanding of the shaping and the form of the products of volcanic action, examines the Moon, will be amazed to find its entire surface covered with examples of such products and will discern upon it all types thereof in the greatest abundance, volcanic basins with and without molfettas, as well as actual volcanoes with open and closed throats, with and without lava flows, with and without molfettas and so on.

Aepinus was aware that the moon has no atmosphere and attributed the excellent visibility of these volcanic features to the absence of erosive effects of wind, rain, and snow. He assumed the few irregularities in the dark regions to be rims of ring structures projecting about the surface of the (water-filled) lunar oceans. And the great ray craters he saw as enormous volcanoes; in fact, he held that Mount Etna on Sicily, the largest active volcano in Europe, if viewed from above, would look like Tycho. In 1778 Sir William Herschel added further detail to the theory of lunar volcanism. He perceived "three volcanoes in different places of the dark part of the new moon. Two of them are either already nearly extinct, or otherwise in a state of going to break out, which perhaps may be decided next lunation. The third shows an actual eruption of fire, or luminous matter." The "eruption" he described as resembling "a small piece of burning charcoal, when it is covered by a very thin coat of white ashes, which frequently adhere to it when it has been some time ignited; and it had a degree of brightness, about as strong as that with which such a coal would be seen to glow in faint daylight." What is more, he added, "All the adjacent parts of the volcanic mountain seemed to be faintly illuminated by the eruption, and were gradually more obscure as they lay at a greater distance from the crater."

Sir William Herschel

From a contemporary perspective it may be difficult to understand why people failed to question that lunar craters were of volcanic origin. Although much larger in scale, the formations, when viewed through early telescopes, indeed seemed similar to the Earth's volcanic craters—especially Vesuvius, which has a ring of hills surrounding it. Illustrations based on observations of the moon under slanting sunlight—with shadows standing out dramatically—showed sharp peaks where in fact there were none. Moreover, eighteenth-century science had no clue how the moon had come about in the first place. Finally, no one in recorded human history has seen a meteorite fall and create a crater. From time to time there had been more or less reliable reports on stones coming from the sky. On March 16, 687, for example, in China during the Chou dynasty, one observer wrote, "In the middle of the night, stars fell like rain." But in the early afternoon of July 26, 1803, at a time when even the existence of meteorites was under debate, something remarkable happened. Around the French village of L'Aigle in Normandy hundreds of people witnessed a meteorite shower of thousands of pieces, which not only eliminated all doubt about their extraterrestrial origin but also confirmed the possibility of impacts. Shortly after, on December 14, 1807, the Weston meteorite of Connecticut was the first fall recorded (by two Yale professors) in America. President Thomas Jefferson supposedly commented, "I would more easily believe that two Yankee professors would lie than that stones would fall from heaven." Still, this possibility was not immediately applied to lunar craters. Sir John Herschel, who had been the victim of the *New York Sun*'s "blue bat people" hoax, still likened lunar craters to the volcanoes known in France and Italy and stressed that "they offer, in short, in its highest perfection, the true volcanic character, as it may be seen in the crater of Vesuvius; and in some of the principal ones, decisive marks of volcanic stratification, arising from successive deposits of ejected matter, may be clearly traced with powerful telescopes."

But doubt was growing. Some scientists were not ready to recognize what Herschel discerned through his telescope. The English popularizer of astronomy Richard A. Proctor (1837–1888), in his book *The Moon,* struggled with the theory

of volcanic origin. He knew that water was necessary for the occurrence of violent eruptions,
and he saw no satisfactory explanation for what
he called "the structure of the great crateriform
mountain-ranges on the moon." While conceding
that at some point in time there may have been
large amounts of water on the moon, he saw no
such possibility in its current state. In addition,
he asked himself how the many smaller craters
surrounding the prominent crater Copernicus,
for example, might be explained. In the 1870s

A nineteenth-
century fantasy
of the moon

Proctor proposed that the lunar craters had been formed by meteors, when the moon was "in a plastic condition." "It may seem, indeed, at a first view, too wild and fanciful an idea to suggest that the multitudinous craters on the moon, and especially the smaller craters revealed in countless numbers when telescopes of high power are employed, have been caused by the plash of meteoric rain, and I should certainly not care to maintain that as the true theory of their origin; yet it must be remembered that no plausible theory has yet been urged respecting this remarkable feature of the moon's surface." Proctor imagined that "under the tremendous heat generated by the downfall, a vast circular region of the moon's surface would be rendered liquid, and that in rapidly solidifying while still traversed by the ring-waves resulting from the downfall, something like the present condition would result."

It sounds like a paradox, but in a convoluted way, more research was needed on Earth to explain the craters of the moon. The geomorphologist Grove Karl Gilbert (1843–1918), who worked for the U.S. Geological Survey, began by studying Coon Mountain (later renamed Meteor Crater), an enormous hole in the sandstone of the Arizona desert. When he compared the volume of the crater and the material on the rim, he concluded erroneously that a meteorite could not possibly have gouged out the crater. Even the meteorite fragments found near the crater could not convince him, and he declared their presence coincidental. We now know that a massive stone weighing 63,000 tons exploded there fifty thousand years ago after crashing at a speed of more than forty thousand miles per hour, about fifty times as fast as a bullet. A decade after the disastrous volcanic eruption on the Indonesian island of Krakatoa in 1883, Gilbert shifted the emphasis of his work to the moon. Ironically, he steadfastly

Barringer crater, Arizona

held that the moon's craters were indeed the result of meteor impacts.

According to a common assumption of the time, meteorites that hit the moon from various inclined angles would have caused oval or "stretched" craters. How, then, could Gilbert account for the craters' uniformly circular contours? In his laboratory he fired clay balls at clay targets under controlled conditions and discovered that the nature of the impact scars depended not only on the angle at which the clay ball hit the target but also on the viscosity of the material and the velocity of the impact. The results, he declared, would account for more or less circular impact holes on the moon's surface.

But it still took time for the new theory to gain credibility, as most scientists in the field found it too radical. The amateur astronomer Ralph B. Baldwin provided further evidence that the moon's craters had resulted from impacts. Scrutinizing photographs of the moon, he had long wondered how the straight valleys or grooves had come about that pointed toward the center of Mare Imbrium. He concluded that they could only have been carved out by massive rock formations during violent explosions. After World War II, he correlated depth-diameter values for lunar craters with holes left by bomb explosions and found similar characteristics. He assembled a list of about fifty terrestrial formations, most of which were subsequently proven to be the results of impact. Baldwin wrote in *The Face of the Moon* (1949), "all observations point to the conclusion that the great majority of the lunar craters were born in gigantic explosions, that these explosions were caused by the impact and sudden halting of great meteorites, and that the main features of the moon's crust were established in the first quarter of its life as a satellite." He also realized that the symmetric nature of the shock wave caused the crater to be circular. While working on his hypothesis, Baldwin must have been aware of the atomic bombings of Hiroshima and Nagasaki, but we can only speculate about the effect this knowledge may have had on him. His book sold poorly but got into the hands of the right scientists and is now commonly regarded as one of the most important books in the history of lunar science.

More recently, Dana Mackenzie has given us a clearer characterization of the nature of such impacts: "It isn't really accurate to think of the meteorite 'digging' a crater; instead, it compresses, fractures, pulverizes, and even melts the rock, and launches it in all directions. These secondary 'bombs' create havoc in their own right. On the moon (though not on Earth) it is easy to spot secondary craters that were created by material ejected from larger impact craters."

Meanwhile, competing theories continued to thrive. In the late 1880s S. E. Peal developed a glaciation theory of the lunar surface. He assumed the maria to be filled with frozen water and the lunar surface to be coated with ice. He imagined that the crust had emitted water vapor, which then condensed around the volcanoes and developed into icy craters. The idea of an icy moon persevered into the twentieth century. In 1925 E. O. Fountain of the British Astronomical Association theorized that insulating layers of meteoric dust protected the moon's crust from the heat of the sun, resulting in a permanent layer of ice just below the surface. The German astronomer and selenographer Philipp Fauth (1867–1941) worked for decades on a detailed but flawed moon atlas. He believed in the so-called glacial cosmogeny or world ice theory—particularly popular during the National Socialist period—which held that the moon's crust was a hundred-mile-deep layer of ice.

Edward G. Davis suggested in an article published in *Popular Astronomy* in 1923 that coral atolls had built up in the lunar seas, and that some of the craters were filled to such an extent that even the central peaks could not be discerned. He calculated an age for the rampart of the crater Copernicus of sixty-eight thousand years. Upheavals of the crust finally had brought an end to life on the moon, leaving only the atolls as remnants of a time long lost. Perhaps the strangest theory was published by the eccentric Spanish engineer Sixto Ocampo in the 1949 *Bulletin of the Argentine Friends of Astronomy Association*—after it had been rejected for publication in his homeland. Ocampo was convinced not only that the moon had been inhabited but that the craters and the lunar rays were created by nuclear blasts detonated by two warring races of beings. The different shapes of the craters, and the existence of peaks in the center of some, according to Ocampo, reflected the use of different kinds of bombs. Finally, Ocampo held that the last great explosions on the moon redounded to Earth, causing the biblical Flood.

That which cannot be seen, named, and classified always presents a challenge, even if its presence is clearly evident. So if the visible craters inspired waves of thought and theory, so much more a source of wonder was the far side of the moon. Until a half-century ago, it resisted both observation and naming. The Danish mathematician Peter Andreas Hansen (1795–1874), twice the recipient of the Royal Astronomical Society's Gold Medal, was convinced that the shape of the moon deviated considerably from spherical and that the center didn't exactly coincide with its center of gravity, because of differences in density. According to Hansen, all water and air had moved to the far side, creating the conditions for organic life to evolve there. He was joined by Sir John Herschel, who believed that the far side might contain an ocean of water. While the essence of Hansen's theory has long since been discarded, he was right in thinking that the center of the moon's gravity is slightly askew.

The far side of the moon is commonly called the "dark" side. But the surface is no darker than the face we're familiar with; the expression reflects only our perception of a surface so long hidden from our view. Just like the near side, the moon's far side experiences two weeks of sunshine followed by two weeks of night. We can see a little less than a fifth of the far side under certain conditions, as the result of slight oscillation or "libration"; the rest is forever invisible from Earth. Scientists and laymen have wondered what the unseen side of the moon would be like, and some writers have turned the mystery into metaphor. According to Mark Twain (1835–1910), "Everyone is a moon and has a dark side which he never shows to anyone." A century or so later, a generation of teenagers puzzled over the hidden meanings of Pink Floyd's immensely successful album *The Dark Side of the Moon* (1973), which focused on darker aspects of human existence.

Eventually the mystery of the far side of the moon fell victim to a combination of human curiosity and technological progress. The Soviet Union began a lunar program in January 1959 that eventually completed twenty missions to the moon by the end of 1970. On October 4, 1959, *Lunik* 3 was launched, carrying two automatic cameras equipped with a special orientation and electronic guidance system that kept the lenses directed toward the moon. On October 7 an automated command caused the film in the cameras to be exposed as the craft was passing about forty thousand miles above the lunar surface. From the perspective of the cameras, the phase of the moon had shifted considerably. On Earth the moon appeared just a few days into its cycle, but from *Lunik*'s position in space it was nearly full. Photocells detected the sunlit side and triggered the photo sequencing. Over the course of forty minutes, a series of shots captured about

70 percent of the moon's far side. The exposed film was developed, fixed, and dried automatically on board the rocket. It took a few more days until *Lunik* was close enough to the Earth to transmit the images of the negatives to the ground stations. The photographs were enhanced by computers to produce a tentative map of the dark side.

The biggest surprise was that the surface of the far side differs significantly from the near side, with many more crater impacts. It has been calculated that maria occupy close to 30 percent of the near side but barely 2 percent of the far side. One theory focused on the gravitational attraction of the Earth, which is a bit stronger on

After the far side of the moon had been photographed by the *Lunik* 3 spacecraft in October 1959, these plastic models were fabricated in the former Soviet Union.

the visible side. The force would have been even greater in the past, when the moon was closer to Earth. The average exposure to sunlight, on the other hand, is about the same on both sides, so this factor could not have been responsible for such a difference. Could the Earth have shielded the moon's visible side from meteorites? Probably not, as the Earth's disk occupies only a tiny part of the moon's sky. Ultimately, scientists ruled out external causes for the asymmetry, reasoning that analogous hemispherical differences in continents and oceans on Earth can be attributed to internal causes. The most likely explanation is that the crust on the far side is thicker, making it harder for molten material from the interior to have flowed to the surface and formed the smooth maria. It is also known that a giant impact struck the near side with such force that it created the two thousand–mile–wide Nearside Megabasin and sent vast amounts of ejecta to the far side.

Once *Lunik* allowed the Soviets to map the far side, they went on a naming orgy, dubbing every visible feature. Thus the Soviets not only put the first man in space, but trumped the United States in other endeavors as well. *Ranger 4,* the United States' first attempt to photograph the far side of the moon, crashed into the surface in 1962 before transmitting any pictures. Still, all was not lost. Later space probes, such as *Ranger 7* (launched July 1964) and especially the Lunar Orbiter program (1966–1967), succeeded in photographing the surface in much greater detail. Finally, in 1968, when *Apollo 8* circled the moon in preparation for the *Apollo 11* landing, the crew aboard were the first to see the far side with human eyes.

CHAPTER NINE

Lunar Choreography

"The higher the moon, the higher the clouds, the finer the weather." "Moon in the north brings cold; moon in the south brings warm and dry." Much traditional weather lore is based on such correlations. Poised at an uneasy intersection of astronomy and popular cosmology, astrometeorological speculations about the moon's effects on the weather are particularly persistent. The London pharmacist Luke Howard (1772–1864), who devised the system still used for classifying clouds, maintained an elaborate record of meteorological observations. He saw a correlation between patterns of barometric fluctuations and the moon's gravitational effect on the atmosphere. Some "lunarists," in turn, thought that the moon created electric or "magnetic" atmospheric disturbances. Robert FitzRoy (1805–1865), twice commander of the *Beagle* (the second time with the young Charles Darwin on board, who commented on FitzRoy's "most unfortunate" temper), later became the head of the British Meteorological Office and championed the idea that both moon and sun exert a pull on the atmosphere. His *The Weather Book: A Practical Meteorology* (1863) includes several references to the moon's impact on the weather. Such phe-

Moon + woods = moon wood?

nomena as "more than usual twinkling of the stars, indistinctness or apparent multiplication of the moon's horns, haloes, 'wind-dogs,' and the rainbow, are more or less significant of increasing wind," FitzRoy claimed, "if not approaching rain, with or without wind." Furthermore, for FitzRoy the moon at the last moment during which the obscure disk is visible "is a sign of bad weather in the temperate zones or middle latitudes (probably because the air is then exceedingly transparent)." Such were highly controversial even at the time. Skeptics of the moon's influence on terrestrial weather point out that a change in weather is always taking place *somewhere* on Earth.

An undeniable point of reference for such assumptions, however, has been the action of the tides, which result from the interplay of the gravitational and centrifugal forces of Earth, moon, and sun. This is the moon's most obvious influence on our world. The daily rise and fall of the ocean water—averaging three feet worldwide—is produced as water is drawn toward the moon and, simultaneously, the Earth's centrifugal force creates a corresponding but smaller bulge on the other side of our planet. When the sun, Earth, and moon line up—whether at the new or full moon—the spring tide occurs in which the moon's impact is augmented by the tidal force of the sun. When the moon is at a right angle to the sun, in turn, tides are at their least extreme, the neap tide.

In fact, some contemporary scientists have found a distinct correlation of earthquakes and lunar phases and have speculated that a slight deformation of the Earth's crust could trigger an earthquake. Geoff Chester, an astronomer and public affairs officer with the U.S. Naval Observatory, has stated, "The same force that raises the 'tides' in the ocean, also raises tides in the [Earth's] crust." The earthquake in Greece and Turkey in fall 1999 occurred after a total solar eclipse, and the tsunami of 2004 happened at the time of the full moon. Although the ability of the Earth's gravity to trigger moonquakes has been confirmed, the converse phenomenon is still doubted by others.

Aboriginal Australians in coastal areas have noted a correlation between the moon's phases and the tides. They once believed that the high tide runs into the moon as it sets into the sea, making it fat and round. Conversely, when the tide is low, water pours from the full moon into the sea below and the moon consequently becomes thin. For travelers from ancient Rome and Greece who were used to the relatively motionless Mediterranean Sea, it was a surprise to experience the strength of Atlantic tides. When, in 55 B.C., Julius Caesar arrived on the coast of Kent, a few miles northeast of the chalky cliffs of Dover, to invade England for the first time, he inadvertently anchored his ships at the peak of a twenty-foot spring tide with a strong following wind. After a short foray inland, he returned to find his fleet stranded high and dry on the tidal flats. His army had to face a brutal attack from the Britons.

Myths and tales abound of correlations between the moon and the natural life on Earth. Farmers once thought that the harvest should occur when the moon was full. Rural folk bundled wheat on the threshing floor "during the moon's age" so that it would dry better; grain brought in during the waxing moon would be moist and soft and might burst during threshing. *The Old Farmer's Almanac,* published annually since 1792, recommends that its readers, "plant flowers and vegetables which bear crops above the ground . . . during the *light* of the Moon; that is, between the day the Moon is new and the day it is full." However, the authors continue, "flowers which bear crops below ground should be planted during the *dark* of the moon; that is, from the day after it is full to the day before it is new again." Some gardeners in France, in contrast, according to Camille Flammarion, attributed to the light of the moon in April and May "an injurious action on the young shoots of plants." "They are confident of having observed that on the nights when the sky is clear the leaves and buds exposed to this light are blighted—that is to say are frozen, although the thermometer in the atmosphere stands at several degrees above zero." Farmers plagued by fleas were told to look for them in the moon's glow, because the tiny creatures were assumed to be drawn by the shimmer and would emerge from the clothes where they were hiding.

Do trees feel the moon's influence? In some areas of central Europe, lumber was cut exclusively in the winter—for example, between

Christmas and Epiphany—and then only according to certain rules. This makes sense because wood is generally dryer during this time of the year, but some people believe that lunar phases influence the quality of wood as well. Sometimes "moon wood"—lumber cut during a certain lunar phase, typically before the new moon—is said to have unique properties, such as resistance to weathering or suitability for use as cookware in a fireplace.

Farming in tune with the moon

Preference for moon wood may sound like superstition, but in fact there is some evidence that it exhibits unique stability. By means of exact measurements, the engineer Ernst Zürcher from the Bern University of Applied Sciences has discovered that wood felled just before a new moon has more water in its cells—and thus is heavier and more stable—than wood felled immediately preceding a full moon. Zürcher attributes this phenomenon not to the moon's attraction but to the tree's water economy, hypothesizing that in its new phase the moon shields the Earth from solar winds that directly influence how water binds in the tree's cells. Zürcher also believes that the light of the full moon is capable of bleaching wood. Experiments by the Swiss Federation of Technology's Department of Forestry Science, in turn, have shown that trees planted during a full moon grow better, giving weight to a piece of folk wisdom, though not all scientists agree with these conclusions.

The daily rise and fall of the oceans' water level governs much of the life on their fringes, where habitats are exposed and submerged every day. Anthony Aveni reminds us in *Empires of Time* that "many feeding and reproductive cycles of organisms that inhabit the ocean receive their input signals from the tidal period, which, in turn, is regulated by the position and appearance of the moon." He speculates that "moon-based cycles may be a reflection of our original ascent from the sea." Aveni looks at a coral from the Devonian period (350 million years ago) and recognizes it as "an archaic month calendar frozen in time": horizontal growth ridges have been deposited during full-moon phases. This detail proves that when the moon was closer to the Earth, our months were shorter, and there were thirteen of them in a year rather than just twelve. More important, we have evidence that organisms have been attuned to the evolving lunar period over hundreds of millions of years.

How, then, do these cycles and rhythms manifest themselves in living beings? When Aristotle was observing sea urchins, he could discern that their ovaries tended to swell during the time of the full moon—a phenomenon that modern scientists studying sea urchins at Santa Catalina Island in California have confirmed. When Columbus reached the New World on October 12, 1492, he and his companions saw a candlelike flicker in the water off the coast of the Bahamas about an hour before the moon rose. The naturalist Lionel Ruttledge Crawshay (1868–1943) conjectured that this light may have been caused by the luminescent Atlantic fireworm. During the moon's last quarter, the female fireworm rises to the water's surface after sundown and emits flashes of light. This spectacle attracts the male worm, which in turn speeds through the water flashing like a firefly. Using range maps for the worm, Crawshay postulated that Columbus's first landfall would have been at Cat Island and not San Salvador, as is commonly assumed.

Today, chronobiologists can demonstrate links between biological processes in animals and daily, seasonal, lunar, and tidal cycles. A number of organisms living in tidal zones are attuned to the pulse of the ocean. The Pacific palolo worm (*Eunice viridis*) found in the coral reefs around both the Samoan and Fiji Islands offers a particularly interesting example. This creature breeds in October and November (spring in the Southern Hemisphere), when the moon is in its last quarter. The Samoans have developed an intricate system of monitoring this process, which includes close observation of the scarlet blossoms of the flame tree (*Erythrina indica*), whose flowers' emergence appears to be synchronized with the palolo's

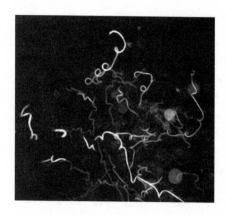
Palolo worms

approach. Shortly after the full moon, when it is low on the western horizon at daybreak, the Samoans know that the palolo feast is just seven days away. At sunrise, the ends of the swarming worms, engorged with sperm or eggs, break off, rise to the surface, and thrash about. Most islanders scoop up a mass and eat the palolo raw as they catch them; others later devour this caviar as a fried delicacy.

Can we establish with certainty what is going on? "Moonlight penetrates quite well in tropical water and these organisms are much more sensitive to light levels than we are," says Paul Brown of the National Park Service, who has repeatedly dived at eighty feet in the middle of the night, with moonlight as his sole illumination. "Palolo tend to be fairly shallow in Samoa and are most common on the reef crest and slightly below, and so are typically in depths of twenty to forty feet." Brown adds: "Even though the breeding process is clearly connected to the lunar cycle, the exact trigger mechanism is still unknown."

The grunion, a small member of the silversides family that lives off the sandy coast of southern California, provides a fascinating example of the attunement of spawning rhythms with the tides. Twice a year, around the end of February and the end of August, thousands of these tiny fish approach the coast. Grunion spawn on the beach chiefly during spring tide—more precisely, during the first three to four nights after a new or full moon in the first two hours following the high tide. Swimming against the flow of the outrushing water until they reach the beach, these small fish transform the beach into a glimmering silver surface.

Accompanied by one or several males, the somewhat larger females dig themselves into the sand with their tails and fins. There they are encircled by the ejaculating males. After they lay their eggs a few centimeters deep in the sand, the females work their way up, loosening themselves to be carried back into the ocean by the retreating waves. It will be two weeks until the next spring tide, but under cover of the sand the young fish develop as if following a secret plan. When the water level again reaches the point of the beach where the eggs were buried, the pounding of the waves lays bare the young grunion, which emerge to be carried with the waves to the ocean.

Other examples from the natural world show a closer affinity of other species with moonlight itself. In the early 1980s scientists discovered a mysterious rite of procreation among corals. Just after the full moon, the researchers observed, corals dissolve in an orgy of reproduction, sowing the waters with trillions of swirling eggs and sperm that merge to form new life. The details long remained obscure—it was unusual, for example, that the breeding phenomenon occurs during different months, though usually in the

Life in the oceans is strongly affected by tides and the moon. Painting by Ernst Haeckel (*Arabian Corals,* 1876)

summer. In 2007 a group of seven scientists from Australia, Israel, and the United States was able to substantiate that the moon regulates this synchronized mass spawning. This synchronicity is possible because the corals contain ancient photosensitive molecules that trigger this unique response to moonlight. "It is an amazing molecule really, triggering the largest spawning event on the planet," says Ove Hoegh-Guldberg from the University of Queensland in Australia. More research is needed to fully understand the secrets of how the corals keep an "eye" on the moon.

How about the moon's influence on land animals? Pliny (A.D. 23–79), known for his interest in the natural world, believed that monkeys are depressed while the moon is waning and happy during the New Moon—but only the ones with tails. According to the Roman writer Aulus Gellius (ca. 130–180), a cat's eyes grow and shrink just as the moon does. The even more imaginative Claudius Aelianus (ca. 170–222) wrote that elephants grazing in the forest when the sickle of the new moon appears gaze at the crescent and wave torn twigs as if begging for the grace, mercy, and protection of the goddess. Aelianus further claimed that the ibis closes its eyes during lunar eclipses until the moon shines again. Pliny, Gellius, and Aelianus, of course, were relating fanciful tales, and later folk wisdom about lunar effects on animal behavior is likewise unreliable. Still, moonlight has been shown to influence the activity of many animals. Contrary to intuition and folklore, some smaller rodents—and even bats, such as the South American fruit bat—are less active than usual on moonlit nights, exhibiting "lunarphobia" or moonlight shyness. The three-striped night monkey or northern owl monkey (*Aotus trivirgatus*), indigenous to Venezuela and parts of Brazil, has the opposite activity pattern: "lunarphilic," during the full moon it is energetic all night.

A key element of lunarphobia or lunarphilia is an animal's place on the food chain. An animal more likely to be caught by predators under bright moonlight, or a hunter that has evolved excellent night vision, is apt to be more active on dark nights. Lunar cycles may also affect the haul-out patterns of harbor seals, when they leave the water for a variety of reasons, including to breed. When the moon approaches its full phase, the seals of Snake Island, British Columbia, stay at sea more often, possibly because the bright moonlight could facilitate the hunting of prey that migrates vertically and thus yield a higher catch. In England, the lapwing usually searches for food during the day and rests at night, but during the winter months the bird looks for food even at night if the moon is full, unless it is covered by dense clouds. And cranes tend to return to their roosts later after sunset if the moon is bright. The skylark (*Alauda arvensis*), a small bird that can be found in Europe and Asia, provides another fascinating example for the influence of moonlight: As a comprehensive study undertaken by French scientists during four fall seasons has shown, the onset of migration of this bird is determined by the phase of the waxing gibbous moon, which facilitates the navigation conditions during their flight to southern Europe and North Africa, from where they return at winter's end. This timing allows the bird to benefit from optimal conditions of illumination for almost a week.

Eric Warrant of the University of Lund in Sweden, who specializes in the vision of animals, has observed the behavior of African dung beetles (*Scarabaeus zambesianus*). He found that this animal is, by virtue of particularly sensitive eyes, able to navigate by the dim polarization pattern of moonlight. The beetle maintains a straighter rolling path under these conditions, the moon acting as an orientation cue.

Dung beetle

As crepuscular animals, active predominantly at dawn or at dusk, wolves are naturally associated with the moon: their wavering howls were once thought to be a message intended for that heavenly body. But the facts provided by canine experts are sobering. Even if a wolf points its faces toward the moon and stars, it is not really howling at the moon. The reason for its posture is purely acoustic — projecting its call upward simply allows the sound to carry farther.

The iconic wolf

Thanks to the high pitch and drawn-out notes, a wolf's howl can travel as far as six miles in the forest and ten miles across unobstructed tundra.

This final example may disappoint some romantic nature lovers, but the fact remains that some rhythms in the animal world are related to the tides and the phases of the moon — that some animals possess internal clocks able to "predict" the lunar cycle. Their lives are indeed affected by the moon — but what about our own?

CHAPTER TEN

Esoteric Practices

If we could travel back in time, we'd arrive at a point when only a fine line separated alchemy, folk medicine, and medical science. The moon figured in all of them. Premodern healing systems are attracting renewed interest today, but two centuries ago or so they were the first choice of much of the populace. The Berlin of the late eighteenth century, for example, was a major center of the Enlightenment, but also the home of a moon doctor, a Mr. Weisleder. By the early 1780s Marcus Herz, one of the most renowned Jewish physicians in the city, decided that this dubious (but irksomely popular) character must be a traitor. Herz scrutinized his rival's practices and recorded his observations in a long piece entitled "Pilgrimage to the moon doctor in Berlin." At five o'clock one summer afternoon, when Herz arrived at the "temple of Asclepius" in a "wretched beer house of the lowest class" in the Jakobstrasse, he found crowds of people from all around the city climbing the back staircase to the building's second floor. There, it was said, Weisleder would soon begin to treat patients of all religions and from every stratum of society. He would cure any and all ailments, mutilations, and physical handicaps, from fractures and hernias to

The Hungarian astronomer Maximilian Hell (1720–1792) thought that he could heal certain ailments by redirecting the forces of the moon.

inflamed eyes and impaired hearing. "When we got inside," Herz wrote, "all I could see was the empty, dirty, and low-ceilinged parlor of a common craftsman. A big table in the middle and a few stools at the wall represented the whole furniture." A rumor spread that the moon doctor had been called away by a princess and was also awaited by the prince, so the crowd would have to wait for five hours. Doctor Herz bit his lip, fearful that laughter might cost him injury, or even his life. He decided to return in a month.

On his next visit, he again found a large crowd waiting for Weisleder; he counted a few hundred, many of them in elegant garb. He pushed his way through the crowd to reach the front of the line, waiting for the second-floor "divinity." Twelve patients were admitted at a time. An assistant named the ailments, one after the other, and the sick came when beckoned. Doctor Herz pretended to have an attack of gout. The moment Weisleder made his appearance, the sick paused for a moment in veneration. Herz was less impressed. This tall, gaunt man, indifferently walking about the room holding a pipe, was a former hosiery knitter with "combed out hair" clinging to his coarse blue frock. "His physiognomy was one of the vilest kind," Herz wrote. The moon doctor used the southern window, while his wife treated female patients at the north window. These charlatans distributed no pills, drops, or bandages; each simply instructed the patient to extend the afflicted body part out the window toward the moon. Weisleder stood still, folded his

hands, and murmured a secret text in a voice too soft for the visitor to discern. To be effective—such was the claim—this procedure had to be repeated on three successive days during the first quarter of the moon; Herz later learned that some patients had to subject themselves to this routine for three or four months. Weisleder claimed to have healed conditions that had been declared incurable by regular doctors, including the fracture of a renowned "Madame N.," who would not have to use a bandage anymore. He demanded no

The curious moon doctor drew large crowds to his house. He made his patients point their ailing body parts in the direction of the moon.

fee for his service, but some patients gave his wife money.

Herz commented sarcastically on the astral medicine and supernatural cures of the moon doctor. He even traced the doctor's ministrations to the digestive problems of the former hosiery knitter, who he assumed had been weakened by his constant sedentary position and suffered from inhibited circulation. The man's condition, Herz speculated, "formidably" produced wind "of a hippocratic kind," which, instead of going its natural course, rose up and transformed the man into a fortuneteller, a visionary, a doctor. Herz would have been thrilled to have observed an actual cure, he wrote, but all he saw was an "abominable appearance," the "sight of debased and befouled humanity." Some in the general public acknowledged that Weisleder's cures were often ineffective; some even ridiculed him. Inexplicably, though, many defended and even patronized him. Herz called the moon doctor's practice a common form of "mischief," the mere thought of which was capable of producing in him "a surging heat" for years to come. Later he learned with satisfaction that Weisleder had to close in the face of undeniable evidence that his procedures produced no effective results.

The moon doctor is representative of a whole tradition of practicing medical astrologers, who believed that there are celestial and lunar influences on particular diseases. Some drew an analogy between the human body and the atmosphere, both of which were thought to be influenced by the moon and sun. The German doctor and healer Franz Anton Mesmer (1734–1815), for example (whence the term *mesmerized*), was convinced that certain psychological symptoms in his patients changed with the moon's phases. Believing that there are "invisible fluids" within the body, he attached magnets to patients' bodies to induce "artificial tides" that would influence those fluids. His practice was somewhat akin to widespread bloodletting, which was meant to lessen the viscosity of bodily fluids.

The belief persisted among many astrological doctors that women were particularly liable to lunar influences. Men were assumed to be reasonable and resistant to natural conditions, but women were closely associated with nature and considered more susceptible to celestial influences. The Dutch anatomist Theodor Kerckring (1640–1693) described a French matron as beautiful and round-faced at the full moon, but when the moon was waning, her face became deformed, her eyes and nose turned on one side. During this phase, she would stay inside, waiting until the moon changed again.

Richard Mead (1673–1754), Physician in Ordinary to Saint Thomas's Hospital, Southwark, seems to have been particularly aware of the conflicting theories of his age. On the one hand, Mead embraced the mechanistic philosophy of Newton, but on the other hand, he clung to belief in celestial influences on particular diseases and to traditional lore linking meteorological conditions to phases of the moon. Reflecting on the loss of reverence for the moon, he lamented that "we can scarcely believe that the present Moon is the same which shone in the days of old. —It is true her aid is sought in navigation; otherwise, she now wanders round the earth for little more than the amusement of juvenile astronomers."

And even Erasmus Darwin, for all his scientific bona fides, wrote in *Zoonomia; or, The Laws of Organic Life* (1794–1796) about the lunar and solar influences on a number of diseases, declaring that over longer periods of time the heavenly bodies could "produce phrensy, canine madness, epilepsy, hysteric pains or cold fits of fever." At the same time, Darwin stressed that "the periods of quiescence and exacerbation in diseases do not always commence at the times of the syzygies [straight-line configurations, producing full and new moons] or quadratures [right-angle alignments, at the first and last quarter] of the moon and sun, or at the times of their passing the zenith or nadir; but as it is probable, that the stimulus of the particles of the circumfluent blood is gradually diminished from the time of the quadratures to that of the syzygies, the quiescence may commence at any hour, when co-operating with other causes of quiescence, it becomes great enough to produce a disease." In his painstaking attempt to prevent returns of diseases, Darwin then asked,

> Do not the cold periods of lunar diseases commence a few hours before the southing of the moon during the vernal and summer months, and before the north-

ing of the moon during the autumnal and winter months? Do not palsies and apoplexies, which occur about the equinoxes, happen a few days before the vernal equinoctial lunation, and after the autumnal one? Are not the periods of those diurnal diseases more obstinate, that commence many hours before the southing or northing of the moon, than of those which commence at those times? Are not those palsies and apoplexies more dangerous which commence many days before the syzygies of the moon, than those which happen at those times?

One can't accuse Darwin of giving the matter too little thought.

Often doctors saw a connection between fever and the moon if only to the extent that fever was thought to "ebb and flow" in parallel with the lunar cycle. In the tropics, where the heat, so medical men tended to believe, relaxed the fibers of the body, this link became particularly evident. Many physicians who worked in British military hospitals in India or the West Indies maintained that the moon has "a more corruptive power" there, in a process comparable to the rapid putrescence of exposed meat and fish. Benjamin Moseley (1742–1819), a member of the Royal College of Physicians of London, recalled that "all the soldiers in the military hospitals in Jamaica, under my care, in dysenteries and intermittents, almost certainly relapsed at the lunar syzygies." Moseley's friend Sir Richard Worseley observed similarly that "afflictions of the skin and eye" in the eastern Mediterranean "fluctuated according to the waxing and waning of the Moon." These doctors often modified their procedures according to the medicine practiced by indigenous peoples, who also tended to believe in the power of the sun and moon.

Times of crisis inspired particularly strong beliefs in lunar influences on human health. When, in 1817, cholera first became endemic in India, it was linked to the influence of the moon. The surgeon Reginald Orton associated the new and full moons and the periods when the moon is closest to the Earth with disease. According to his experience, it was at such times that the majority of cholera deaths occurred. Such explanations, incidentally, were not generally held by the medical profession in Orton's British homeland.

Although the nineteenth century was marked by many industrial and intellectual advancements, as more people moved into cities and lost touch with nature and the nighttime sky, some popular interest in occult knowledge nonetheless survived. Often the moon was associated with evil forces. In fact, a number of secret societies embraced magic rites that celebrated supposed supernatural powers. In various cultures the moon is accorded a place in regional medicinal systems unrelated to the tradition of Western medicine. Acupuncture treatment, for example, is particularly recommended during the day before and after the full moon. The Chinese principle of yin represents both the Earth and the moon, while yang stands for the sun. During the time of the full moon the human body is believed to be in a state of excess; doctors, as a general therapeutic principle, aim at restoring a state of balance. Ayurveda, an indigenous tradition in use for several thousand years in India, is recognized by the World Health Organization as a sophisticated medicinal system. According to its precepts, certain seasons and lunar phases enhance the efficacy of plants used to treat certain ailments. Some traditional medical systems in Africa and South America also recognize the influence of heavenly bodies; one common belief is that herbs collected during the full moon have maximum curative power.

A French tarot card

Belief in occult wisdom and rituals involving the moon thrive in some Western societies as well, as if the Enlightenment had merely pulled a thin veil over millennia of traditional lore about the moon and its alleged involvement with terrestrial life. In fact, an underside of irrational thought has always counterbalanced rationalism, resisting the dramatic change wrought by science and technology in humanity's worldview. Full-moon rituals are common among New Age, New Pagan, and occult groups in some Western countries. Adherents to the Wicca movement, for example, perform rituals to "truly bathe in the touch of the Divine." Typically, the moon of each month is associated with a characteristic image:

the storm moon of March, the wind moon of April, or May's flower moon. The specific ritual homage to the current "magic moon" varies, but the rite is usually performed on an altar, with candles, burning incense, and spiritual music.

In spring, when "the frigid cold of winter has been replaced by the promise of new life and growth," offering, "a chance at fertility and abundance, rebirth and regrowth," the altar will be decorated with fresh cuttings from the garden or packets of seeds to celebrate the arrival of the full moon. Harking back to a time when the well or spring was seen as a sacred and holy place, participants bring small bowls of water and form a circle. The ritual is led by a priest or a priestess. Facing the moon, the celebrant holds the bowl of water to the sky and says, "The moon is high above us, giving us light in the dark. She illuminates our world, our souls, our minds. Like the ever-moving tides, she is constant yet changing. She moves the water with her cycles, and it nourishes us and brings us life. With the divine energy of this sacred element, we create this sacred space." The priestess then dips the cut flower into the water and, while walking a circle, sprinkles water on the ground with the petals of the flower. Next she approaches the participants, carrying a large bowl; each person pours water from a small bowl into the larger one, telling where the water was collected—for example, "This is water from the creek behind my grandmother's farm." When everyone has contributed water, the priestess uses the cut flower once more, stirring and blending the water in the large bowl with the stem of the flower, then listening to "the voice of the moon" above. As the forehead of each participant is anointed with the blended water and an individual symbol, such as a pentagram or a triple moon image, the priestess adds, "May the light and wisdom of the moon guide you through the coming cycle." Finally, she reminds them to "consider how our bodies and spirits ebb with the tide, and how we connect to the cycles of water and of the Moon."

What should we make of such a full-moon ritual? Obviously, the followers of these cults are not satisfied with the worldview contemporary science has to offer. Wiccan practice is at odds with the doctrines of the scientific establishment, but it should make us consider how we deal with the natural world. Even if we may not share such beliefs, they merit respect as attempts to restore a lost magic and faith in nature and the moon — efforts that may ultimately address the deeply felt human need to feel more connected with a world whose forces and mechanisms often evade our grasp.

While a critical examination of the premises of folk wisdom is always worthwhile, the significance of premodern belief systems in Western societies should never be underestimated. It is worth asking ourselves why, for example, in contemporary Denmark fewer nurses and medical doctors graduate from universities than the number of graduates from schools of alternative healing.

CHAPTER ELEVEN

Spurious Correspondences

No one can deny the existence of certain atmospheric influences on the human body. Sunlight, the duration and intensity of which directly correspond with the seasons, not only burns us if we get too much of it, it can affect our moods. Thunder and lightning can arouse strong fears and end lives. Extreme changes in atmospheric pressure can cause lungs to collapse. Correlations have been shown between the seasons and the birth and death rates. But does the moon actually influence the physiology of humans? If so, how much?

A whole body of folklore explores the moon's influence on various phenomena of human life. According to German folk wisdom from the 1800s, for example, those who worked by moonlight risked being slapped in the face by an invisible hand or even going blind. Use of the spinning wheel at night, a common household activity among poor people before the Industrial Revolution, was thought to yield worthless cloth—or, in an extreme version of the superstition, the threads that would be woven into a rope that would end up around a relative's neck. Hanging laundry to dry in the moonlight was believed to wear out the fabric or cause it to absorb poisonous night dew. Moonlight was not to fall

The legendary gambler in Gold Rush California Eléonore Alphonsine Dumant (also known as Madame Moustache) was not shy about using a gun at night onto the marital bed, and children were not to be conceived by its glow, because such pregnancies were fated to end in miscarriage or the birth of an insane child. People who urinated in the moon's direction were at risk of developing a badly swollen eye, and those vomiting in its direction could expect a rash of the mouth. Dancing under the moonlight, especially in a tight embrace, was considered dangerous because Earth's crust was thought to be particularly thin under moonlit conditions. Moreover, the dancers' tapping feet could create vibrations that would lure ghosts from beneath the surface.

Of course such superstitions are hardly limited to European traditions. They can be found in various cultures around the globe. According to a saying from the Philippines, bathing during a full moon results in insanity—and during a new moon, death. In some yogic traditions, both full- and new-moon days are observed as holidays because of the exceptional energy released by the relative positions of sun and Earth. The energy of the full moon is believed to correspond to the end of inhalation. An expansive and upward-moving force, it makes practitioners feel emotional rather than well grounded and makes exercise difficult. In contemporary India some doctors avoid performing surgery under a waxing moon, which they believe encourages scarring. And there are other lunar taboos. For example, some cultures consider the time of the new moon unfavorable for undertaking business.

In Western cultures, some have claimed that

humans can shift shapes to become werewolves, wolves, or other animals, demonstrating violently aggressive behavior and sometimes devouring raw meat. Supposedly, a summer night with a full moon shining directly on the subject's face is particularly conducive to this so-called lycanthropy. In 1977 Harvey A. Rosenstock and Kevin R. Vincent described in the *American Journal of Psychiatry* the case of a forty-nine-year-old married woman who, at the full moon, imagined herself to be a wolf. She became sexually aroused, experienced homosexual impulses, suffered "irresistible zoophilic urges and masturbatory compulsions." This profile may sound odd to contemporary ears and even make us smirk. Today, the extremely rare medical condition of lycanthropy is believed to be caused chiefly by schizophrenia or bipolar disorder—serious scholars rarely if ever link it with the moon. Yet one residual lunar mythology may be connected with a medieval belief in werewolves. Some people claim it's a good idea to trim or cut hair during the full moon if you want your hair to grow thicker and fuller; if cut at the new moon, it might grow back too fast.

We have seen several examples of medical astrology positing dubious correlations between human ailments and lunar phases. While modern science has found little evidence of any causal relationship, some people persist in seeing connections, particularly in the realm of mental health. At the notorious Bethlehem or Bedlam Hospital in London, until 1808 inmates were chained and flogged during certain lunar phases to prevent

"Lunacy"

violence. Joseph Daquin (1732–1815), a Swiss doctor considered a pioneer of psychiatry, believed that the inmates of a mental asylum suffered literal lunacy: the condition was particularly evident during nights with a full moon. Although these observations were not "scientific" in the modern sense, they reflect some strong impression the writer must have experienced. The so-called Lunacy Act of 1842 characterized a lunatic as a person "afflicted with a period of fatuity in the period following after the full moon." The impetus to define a medical condition and pass a law based on it speaks to the conviction that must have reigned regarding the moon's effects on man.

Even our language reflects such superstitions. The erroneous connection of the moon with insanity is reflected in various Indo-Germanic languages—*lunatic, loony,* or *moonstruck* in English, *lunatico* in Italian, *lunático* in Spanish. The Italian idiom *avere la luna,* like the French *avoir des lunes,* means to experience a nervous breakdown. The German *Laune* for *mood* is etymologically connected to *luna,* implying that a changing state is influenced by a phase of the moon. Shakespeare was pointing toward something along these lines in *As You Like It* when Rosalind calls a tormented, confused, hopelessly enamored boy, "but a moonish youth." And while we may recognize today that these connections are false, surely the lines between fact and fiction are not so neat.

In some cases these erroneous ideas have become so enmeshed with our thinking that we hold them to be common sense. For example, conventional wisdom accepts that a full moon interferes with the phases of deep sleep. But is this really the case? Someone who has spent a sleepless night may realize afterward that there was a full moon, and some people have claimed that the light of the

full moon wakes them. Light has been proven to influence activity patterns during sleep, but the first reaction to excessive light would probably be to turn over or cover one's face.

Many people think that they are influenced by the moon.

Many such cases may simply be instances of self-fulfilling prophecy. To a certain extent, something can become reality if we expect it to happen. Applying this logic to case at hand, one predisposed to believe that the full moon interferes with sleep might selectively discard the memories of sleepless nights with no full moon. Could it be *awareness* of a full moon might interfere with the sleep of someone who believes in the connection? Sleeplessness may have many causes, and could even indicate a serious health condition, so for some, blaming the full moon might provide a welcome and comforting explanation—though possibly a dangerous one.

Various studies across nationalities and cultures have sought a correlation between the moon's phases and sleep patterns. In the vast majority, no such correlation could be established, but a few exceptions suggest that a minor connection may exist. A study in 2006 in a suburban area of Switzerland involved thirty-one volunteers over six weeks, including two full moons. Researchers found that subjective sleep duration varied somewhat with the lunar cycle, from six hours, forty-one minutes at the full moon to seven hours at the new moon. There was also evidence that the amount of fatigue the subjects felt in the morning was associated with the moon's phase, with participants feeling more tired when the moon was full. Then again, those taking part in the study had to be aware of the moon's current phase, which may have led them to expect that they would sleep better or worse. The results can scarcely be called conclusive.

Another phenomenon often associated with the full moon—lack of evidence notwithstanding—is sleepwalking, or lunatism. Sleepwalking is a complication in which the sufferer "gets stuck" in his sleep, usually during the first third of the sleep cycle. He may not be fully conscious but is able to perform routine tasks, such as getting up and dressing. Usually, the sleepwalker has no memory of his nocturnal activities the next day. In extreme manifestations of lunatism, the affected person arises and, in an attempt to get oriented, is attracted to sources of light. In the stereotypical manifestation of this phenomenon, the somnambulist may venture onto a rooftop in a futile attempt to get closer to light—often, the light of the moon.

In an earlier time, before artificial light became ubiquitous, moonlight may indeed have been responsible for the disruption of sleep—just as today sleep can be disturbed if someone leaves a lamp on at night or has a streetlight outside an unshaded window. And longer spells of continuous sleep disturbance can certainly have detrimental effects on psychological well-being. But because the moon is a relatively minor source of illumination for most people today, it seems unlikely to have much effect on sleep. Charles Raison at Emory University goes so far as to hypothesize that the persistence of the belief in "the moon's power over the mind" may in fact be a "cultural fossil"—"a memory of an actual effect that no longer obtains." He adds: "Perhaps the moon once had a power over the functioning of the brain that it has since lost."

Might other forces emanating from the moon influence us? Could the moon's gravitational pull—which moves the vast oceans—have a direct effect on humans, whose bodies, after all, are more than three-quarters water? The psychologist Arnold L. Lieber made this claim in his popular book *The Lunar Effect* (1978), but it has been disputed by contemporary scientists. The astronomer and cosmologist George O. Abell once formulated the curious calculation that a mosquito exerts a greater gravitational pull on your shoulder than does the moon. Others have said that the pull would be less than that of a wall of a building at a distance of six inches. One critical consideration is that the water of the oceans is unbounded, whereas the water in the human body is contained within our tissues, which shield it from the moon's tidal forces.

Another common belief, widespread since the eighteenth century, is that a correlation exists between certain lunar phases, the menstrual cycle, and fertility. The word *menstruation,* of course, derives from *mensis,* meaning *month,* which is again related to *moon.* But etymology and causality are two different things. A decisive disparity exists. Although the average menstrual cycle is 28 days, with some variation from woman to woman and month to month, the length of the lunar month is consistently 29.53 days and therefore out of step by a significant margin. The similarity of the time spans appears to be mere coincidence. Nor has a convincing argument been posited why natural selection should favor a method of reproduction based on the lunar month. Furthermore, while the estrous cycle of most primates approximates the human menstrual cycle—ranging between 25 and 35 days—one would expect that other mammals would synchronize with the lunar cycle as well. On the contrary, rats have an estrous cycle of 4 or 5 days, elephants of 16 weeks. Some research, though, has shown some interesting correlations. A study of 826 women, for example, found that menstruation was relatively common around the new moon. Another study of 140,000 births in New York City showed a slight increase of fertility during the last quarter of the moon. On the other hand, a study of a Dogon village in Mali, West Africa, showed no lunar influence on menstruation, even though these villagers don't have electric lightning and spend most nights outside, exposed to moonlight.

Do women living together menstruate in sync with one another? This phenomenon has been observed among some women in some settings, but serious scientific investigations have not established any connection with lunar influences. Is there really a rhythmic, cosmic, or lunar cycle that has been disturbed or masked by civilization, specifically by indoor electric lighting? Such a theory might seem plausible until one considers that many other mammals on the planet have reproductive cycles that are not synchronized with the lunar cycle.

Perhaps the clearest conclusion to be drawn from this range of evidence is that, by virtue of its sheer ubiquity, conventional wisdom dies hard. We see, for instance, that a single study that finds a minor correlation between phases of the moon and this or that phenomenon usually receives a tremendous amount of attention, as if validating our hunch. For instance, Italian bio-mathematicians and gynecologists have studied the relationship between lunar position and the day of delivery. Trying to avoid what they perceived as pitfalls in earlier research, they focused on all the spontaneous, full-term deliveries occurring at one hospital in Fano in the Marche region during a three-year period in the early 1990s. Although they observed a "statistically significant" connection, they found it "too weak to allow for predictions regarding the days with the highest frequency of deliveries." But exactly such a predictive power would be necessary to optimize the resources of a hospital. In short, even the most rigorous scientific studies, able to lay bare phenomena imperceptible to a single individual, have revealed only effects of the moon so weak as to be irrelevant to our daily lives. Still, belief lives on. Another study with a similar focus on births and birth complications, carried out in North Carolina, considered more than 564,000 births between 1997 and 2001 and could establish no connection between frequency of deliveries and lunar phase.

Maybe we can simply blame the media. Any entertaining anecdote about some alleged lunar effect, no matter whether it stands up to scrutiny or not, is sure to arouse public interest and, therefore, to be widely reported. Studies that yield unspectacular results attract no media coverage. In 1997, for example, J. M. Gutiérrez-García and F. Tusell studied almost nine hundred suicides reported by the Anatomical Forensic Institute of Madrid but could establish no significant relationship between the lunar cycle and the suicide rate. Ivan Kelly, James Rotton, and Roger Culver, psychologists from the United States and Canada, examined more than one hundred studies on lunar effects to get a clearer picture. They analyzed numerous phenomena—homicides, traffic accidents, crisis calls to police or fire stations, suicides, kidnappings, stabbings, and various other examples of unusual behavior—and concluded that the studies failed to show a reliable correlation with the full moon or any other of the moon's phases. Remember the splash those reports made in the papers and on the Internet? It's just not much of a story if during the full moon . . . nothing unusual happens!

The belief in lunar influence on our lives has a long pedigree. Given its constant perpetuation in books of pseudoscience and other popular publications, it's not likely to give way anytime soon to a more realistic sense of the moon's power. But scientists and intelligent laypeople alike must continue to question assumptions passed down the generations. Russell G. Foster and Leon Kreitzman have provided many insights into the ways biological rhythms affect life on Earth, but they remind us: "Despite the persistent belief that our mental health and a multitude of other behaviours can be modulated by the moon's phase, there is no solid evidence that the moon can influence our biology." Or, to put it the other way around, as the writer of popular astronomy Bob Berman does, "If all the Moon's alleged powers were real, we'd be a species of lunatics controlled from outer space."

CHAPTER TWELVE

Visions of the Moon

We have seen that mapping the moon began as soon as telescopes provided a closer look. By the middle of the nineteenth century, at the time when Jules Verne was writing his books about lunar travels, moon maps had become more sophisticated and detailed than early ones, but they still failed to create a truly vivid image of the moon. How could anyone give an impression of what it would be like to actually stand on the moon rather than merely to look at it from afar? How could the still invisible and unknown be made visible? Paradoxically, to achieve such verisimilitude required creation of an illusion.

Driven by such an ambition was the Scotsman James Nasmyth (1808–1890). After making a fortune as a manufacturer, Nasmyth had retired to devote himself to astronomy. In the early 1840s he constructed highly detailed and three-dimensional models of the moon's surface out of plaster, then photographed the models. Much of Nasmyth's fascination with the moon came not from hours spent behind a telescope, as one might conjecture, but rather from a trip to Mount Vesuvius near Naples, where he "became in a manner familiar with the vast variety of those distinct manifestations of volcanic action, which at some inconceivably remote period had

The pogo stick was supposed to be powered by a small rocket and designed to permit a hopping range of twenty-four miles.

produced these wonderful features and details of the Moon's surface." From the top of the volcano he could look down into the "pit from which the clouds of steam are vomited forth." To produce the models for his photos, he also drew on the expertise of his father, the landscape painter Alexander Nasmyth.

After thirty years of research and work, James Nasmyth published *The Moon: Considered as a Planet, a World, and a Satellite*. Printed in 1874 by John Murray, who was also Charles Darwin's publisher, Nasmyth's impressionistic photographs presented an artificial moon, the lunar landscape strangely illuminated. Prepared using the arduous Woodburytype process, with a relief mold made of the image in lead, these images were regarded as technically superior to daguerreotypes of the real moon. They provided the viewer with a thrill typical of special effects. Even *Nature* magazine, with its reputation as a guardian of rational science, endorsed the book, a reviewer writing that "no more striking or truthful representations of natural objects have ever been laid before his readers by any students of science."

Shortly before Nasmyth's publication, the American clergyman Edward Everett Hale (1822–1909), in two short stories published in the *Atlantic Monthly* in 1869 and 1870, formulated the idea of an artificial moon. The Brick Moon, measuring about two hundred feet in diameter, was to be hurled into space by flywheels rotating in opposite directions. In Hale's fictional account, the satellite failed to assume its scheduled orbit when launched and seemed lost, but it later reappeared five thousand miles above the Earth, populated by self-sufficient lunarians equipped with a gigantic telescopic lens made of ice. Of course, there is hardly a material less appropriate than brick for a spacecraft, but all peculiarities of this fantastic story aside, we may still consider it an early precursor to the idea of a space station.

In the early 1850s, while Nasmyth was at work on his models, a hemispherical model of the moon's visible side was constructed by Thomas Dickert in Bonn, Germany. The model was about twenty feet in diameter, with fea-

Relief bei Mädlers Halbkugel bei Mondviertel, ausgeführt von Th. Dickert in Bonn.

tures partly based on Johann Heinrich von Mädler's map. Dickert's model was transported to the United States and ended up in Chicago's Field Museum, where it helped to popularize interest in the moon's makeup and astronomy in general. In 1898 Elias Colbert, former director of the Dearborn Observatory in Evanston, wrote in the *Chicago Tribune:* "A close examination of the mammoth Moon enables me to testify to its accuracy as a real model and not mere show. It reproduces with marvelous fidelity the features of the lunar surface and really exhibits a great deal more detail than would be noticed by the unpracticed eye in viewing her through a first-class telescope." The mammoth moon can be seen as an early precursor to the space shows at the world's fairs of the twentieth century. This specimen can still be seen at the Science Central Museum in Fort Wayne, Indiana.

In 1925 the optician and geophysicist Frederick Eugene Wright (1877–1953) established in Washington, D.C., a "Committee on the Study of Physical Features of the Surface of the Moon." Wright's interests in both the moon and the underwater seascape of the Caribbean led him to be dubbed "Moon-Man" and "the modern Jules Verne." To facilitate the mapping and interpretation of the lunar features, he projected negatives of the moon onto one-foot chemical flasks that had been coated with photographic emulsion. They extended over a distance of 130 feet in an underground tunnel. This so-called Wright's Sphere was an early example of rectified photography, designed to restore the original spherical spacing of features, with each assuming its relative position on the globe.

Later in the century, visual portrayals of the moon fell more to artists in the popular, even the pulp, press. A subgenre of space art begins with the rise in the 1920s of science fiction magazines and popular science magazines such as *Science Wonder Stories* and extends through the 1950s and 1960s, the golden age of space enthusiasm, with certain techniques typical of each period. Some consider the American Chesley Bonestell (1887–1987), whose space art was first featured in *Life* in 1944, as the father of astronomical art, but artists such as Lucien Rudaux of France, Klaus Bürgle of Germany, Ralph A. Smith of England, and the American Fred Freeman also made seminal imaginative contributions. All science fiction artists creatively confronted the challenge of

making unseen worlds seem real to a demanding public—believability being more important than accuracy, which, until late in the era, was a mystery anyway.

These artists, of course, worked in tandem with writers who were painting word pictures of the unknown worlds. And despite the strong competition from the planets—above all Mars— the lunar theme continued to be eagerly taken up by writers of science fiction, with the turn of the twentieth century marking a general shift to the dark and dystopic. The moon imagined by the British author Herbert George Wells in *The First Men in the Moon* (1901) is an aging, porous body interspersed with large interconnected caves. The inner moon is full of air and contains the habitats of the lunar beings, who dare to come to the surface only occasionally. Two British travelers

launched from Earth with the help of a substance with antigravity properties experience a dramatic landing. After their space vehicle rolls into the depth of a lunar crater, they are captured by ant-like moon creatures. These savants have balloon-like heads, but their limbs have completely degenerated; their race is doomed. In keeping with its elegiac tone throughout, the book ends with only one traveler managing to escape and return to Earth.

Inevitably, science fiction was soon matched with the new technical possibilities of moving

Attack by insectlike lunarians in Wells's *The First Men in the Moon*

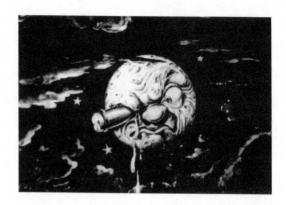

The rough landing of the rocket in the right eye of the moon—a memorable scene from Georges Méliès's *A Trip to the Moon*

images. Georges Méliès, a French actor, theater owner, producer, and director, became the leader of the newly emerging French film industry. His fourteen-minute-long film *A Trip to the Moon,* based on the novels of Jules Verne and H. G. Wells (1866–1946), premiered in 1902 and was soon acclaimed worldwide. The rocket piloted by the long-bearded Professor Barbenfouillis (played by Méliès himself) and five professors lands in the lunar eye, and the visitors are flung to the ground by a violent explosion of unknown cause. They then settle in for the night, only to be awakened by snowfall. Retreating into an area under the surface, they discover strange rock formations, waterfalls, and gigantic mushrooms. But the lunar subsurface turns out to be a kingdom, whose strange beings attack the intruders. When the besieged Barbenfouillis strikes a creature with the tip of his umbrella, it explodes in a cloud of smoke. When more Selenites appear, the visitors are caught, bound, and presented to the king, but they manage to break free and defeat the king.

Ultimately, the travelers flee to their rocket, now perched on the edge of a cliff. It is launched when the passengers' weight tips it over the edge and, with one Selenite on the tail of the rocket, it falls toward Earth. After a brief journey through space, they manage to alight safely on the surface of one of Earth's oceans. Today, this film seems a simple and laughable series of static tableaux, but it remains a real pleasure.

About the same time in New York, the idea of a flight to the moon was combined with panoramas and movement to produce an "electroscenic mechanical illusion." A Trip to the Moon in the Luna Park of Coney Island in Brooklyn was a collectively experienced spectacle. The visitors entered the cigar-shaped and brightly illuminated ship *Luna IV*, which soon began to move back and forth, its large batlike wings moving up and down. Facades whizzing past in the background created the illusion that the ship was moving: from Manhattan to Niagara Falls and then up into space; wind-producing fans enhanced the impression of flight. When the goal of the trip came into view, the wings began to move faster. The ship had encountered a thunderstorm above the lunar surface and just missed landing in an extinct volcano. The visitors were asked to disembark and inspect the rocky landscape with its strange mushroomlike vegetation. Here they encountered midget moon men with rows of long spikes on their backs, who escorted them through stalactite-laden caverns and across chasms spanned by spiderweb-like bridges, all the while singing "My Sweetheart's the Man in the Moon." When they arrived at the fantastic palace of the Selenite king, moon maidens performed a dance for the visitors and offered them bites of green cheese. The guests left the attraction through the mouth of a moon calf and exited into the daylight.

The moon was no source of innocent amusement for Karl-August von Laffert, who foresaw a catastrophe of biblical dimensions in his novel *The Demise of the Luna* (1921): the moon falls to Earth, creating an enormous tidal wave and nearly eradicating humanity. In a time when Europe and Asia have combined to form a kingdom with Eurasiapolis as its capital, the inhabitants have for years awaited a lunar cataclysm, a calamity understood as an inevitable fate. Because of previous volcanic activity, houses on Earth are constructed on a system of clips or are hung from elastic straps for safety. A system of astronomical and meteorological stations has been established across the planet, including one in the Himalayas. After shedding its "icy shell," the moon becomes red, because its metallic core has been exposed, and a second Great Flood ensues. Attempts to transport part of humanity to another planet fail because Earth is the only hospitable place in the solar system to receive "the divine fruit of human seed." All the resources of science and technology are devoted to saving humans, plants, and animals, "to bring them over to the imminent alluvium." A rescue ship has been built—a sort of Noah's Ark where animals have been selected by naturalists—as well as a number of large airplanes for selected humans. Earth and moon finally come together at "flat, almost tangential paths."

Scene from Fritz Lang's film *By Rocket to the Moon*

Compared with Laffert's apocalyptic vision, many of the lunar fictions of the era's movies seem restrained. Most focused on the idea of humans traveling to the moon rather than the moon coming to Earth. The short film *A Trip to the Moon* almost has the feel of a fairy tale. The German filmmaker Fritz Lang, though, who had set the standard for dystopic futurism with *Metropolis* (1927), was not known for restraint. His 1929 film *Frau im Mond* (translated literally in Britain as *Woman in the Moon,* but released in the United States as *By Rocket to the Moon*), based on a novel by his wife at the time, Thea von Harbou (1888–1954), combined melodrama with scientific speculation.

Wolf Helius and his assistant Hans Windegger, enthusiasts of space travel, team up with a Professor Manfeldt, who expects to find water, oxygen, and gold on the moon's far side. Another member of the group is Friede, the former girlfriend of Wolf and now the fiancée of Hans. Helius has been blackmailed into including on the trip the gangster Walt Turner, who has stolen precious material from the scientist. If Helius doesn't cooperate, Turner will sabotage his rocket. Finally, after taking off from a water basin, they discover a sixth passenger, a young stowaway named Gustav. When gold is discovered on the moon, Manfeldt and Turner fight, and the professor falls to his death in a crevice. Turner is fatally shot trying to hijack the rocket. When it is time to leave, a shortage of oxygen on the ship condemns one passenger to stay behind. Windegger draws the short straw, but Helius chooses to let him return in the company of his beloved Friede. Upon takeoff, with the boy assuming the role of the pilot, it becomes clear that both Friede and Windegger have remained on the moon, where, in the final shot, they embrace passionately.

Albert Einstein was among the guests at the premiere of Lang's last silent film, which the press celebrated as "a miracle becoming reality." That judgment may not have held up over the years, but in one important respect, Lang's science fiction inspired reality. The film popularized the launch "countdown." One title card read, "20 Seconds to go — lie still — Take a deep breath," and ever-smaller numbers built suspense toward the climactic liftoff.

Forty wagons of sand were brought in from the shore of the Baltic Sea for the artificial moonscape created for the film in the legendary film studios of Babelsberg near Potsdam and Berlin. It's still surprising to see the similarities between Lang's portrayal and the first flight to the moon forty years later: the appearance of the rocket, its flight, even the separation of the capsule. One of the most fascinating similarities is evident when the spaceship orbits the moon. As the lunar surface passes by, we are aware of the extreme curvature of the lunar globe, so much smaller that Earth. The film transcended its medium, popularizing the idea of rocket science, an effect that cannot be underestimated. Indeed, the shots of the rocket in Lang's film were so convincing that shortly after the Nazis came to power in Germany, they banned the film as a security threat to the rocket development program.

Encouraged by the inspiration of science fiction novels and movies, the fantastic idea of colonizing the moon began to be taken seriously. The New York World's Fair of 1939 and 1940, attended by forty-five million people, presented human flight as the culmination of human technological achievement. The fair's World of Tomorrow simulated an American landscape of 1960 as seen from an airplane, and the Rocketport of the Future imagined a trip from New York to London that would take only one hour. Only civilian applications of future flight technology were heralded at the fair, but its military uses were soon to become reality. Spatial and temporal limits were suspended as never before in the exhibits, but the moon was still a distant vision. By 1962, though, at the Seattle World's Fair, NASA had an exhibit on rocketry, satellites, and human space flight.

The picture began to change soon after World War II, when space began to be promoted as a strategically important issue in the new nuclear age. As early as September 1946, a year after the explosion of the atomic bombs in Japan, *Collier's* magazine ran a story by George Edward Pendray about the moon's colonization. Without providing supporting evidence, he warned that

> rockets only a little faster than the German V-2s could bombard the Earth from the moon. With the aid of a suitable guiding device, such rockets could hit any city on the globe with devastating effect. A return attack from the Earth would require rockets many times more powerful to carry the same pay load of destruction; and they would, moreover, have to be launched under much more adverse conditions for hitting a small target, such as the moon colony. So far as sovereign power is concerned, therefore, the moon's control in the interplanetary world of the atomic future could mean military control of our whole portion of the solar system.

Even though the factual basis for Pendray's thesis was feeble, many readers probably accepted that the moon would be an ideal fortress and military base in a nuclear war. In an October 1948 article, *Collier's* carried the argument further by propagating a "Rocket Blitz from the Moon." Rockets were shown rising out of lunar craters and resulting nuclear fireballs spreading across New York City. Many doomsday scenarios devised during the postwar era had their origins in the sky; UFOs sightings became common at about the same time.

We have to remember that a flight to the moon was not yet an event the general public could envision clearly. The interplanetary epic *Destination Moon* by George Pal (1950) was the first American science fiction movie to take up the topic seriously, staging the trip to the moon as "the greatest challenge ever heard in American industry." The film is not particularly suspenseful, and its dialogue sounds rather stiff to contemporary ears, but its rational treatment of its subject matter was revolutionary. *Destination Moon* includes an inventive comic-strip sequence dismissing the "comic book stuff" characteristic of most depictions of moon flight, and provides a primer on the science of rockets, the effects of gravitational pull between Earth and moon, and the logistics of getting back "from this piece of cheese." During the flight the astronauts experience severe pain, their

First men on the moon in the movie *Destination Moon*

faces are distorted, and they suffer from space sickness. The film also serves as a document of the Cold War. Spurred by fear that other forces plan to go to the moon, and that the United States might be attacked from outer space, the "race" is on. The spacecraft is financed and manufactured by U.S. private industry, before the U. S. government buys the technology.

A few years after its more hysterical ventures into lunar speculation, *Collier's* began in 1952 a series of articles to advocate for space travel as an inevitable human endeavor. The spectacular and influential series predicted a successful manned expedition to the moon within twenty-five years. Wernher von Braun (1912–1977) theorized that three rockets would be necessary to reach space, and that the crew would number no fewer than fifty. The designated vehicle would be clumsy, weigh fourteen million pounds, and include various tanks and such elements as thermal protection shields and a base frame for the landing, but a streamlined rocket wouldn't be required for moving through airless space. The *Collier's* articles also made a passionate plea for active science once we arrived on the moon. Artificially triggered moonquakes, for example, might reveal insights about its inner composition.

When lunarians made it into fiction at all at this time, the portrayal was even more ironic than in the nineteenth century, when at least some serious scientists still clung to the possibility of lunar life. A character in Kenneth Heuer's *Men of Other Planets* (1950) posits that "the really fash-

ionable people on the Moon . . . probably live on the side where they can see the Earth." Heuer's explorers make a daring expedition and find that the far side is covered by forest, whose floor comprises a deep layer of ash. The inhabitants are vile. They appear to be digging holes to plant a garden, but

> they set nothing in the holes; and having
> covered them with sticks and ashes, they
> go a little distance off and hide behind
> the bushes, watching; and we notice that
> as each walks he puts his foot down care-

A series of articles published in *Collier's* magazine between March 1952 and April 1954 introduced Americans to the possibilities of a trip to the moon.

fully, constantly on the alert as to where he treads. We should like to go and work with these men; and not looking where we step, we sink suddenly into one of the holes. And eyes gleam from behind the bushes, and we hear the men laugh in the dark.

An equally wry portrayal of lunar life is Arthur Hilton's *Cat-Women of the Moon* (1953), a low-budget science fiction film originally released in 3D, about an expedition to the moon that finds a group of women in a lunar cave, where some breathable atmosphere remains. They are the last remnants of a human civilization on the moon, and they haven't seen any men for centuries. Grandiosely promoted as "the most startling picture of the century," the film revolves around the ultimately failed attempt of the cat-women to steal the spaceship and return to Earth.

Even our amusement parks took a more sensible stance by this time. Disneyland enlisted the cooperation of Trans World Airlines for its Rocket to the Moon experience of 1955. Von Braun himself contributed advice the attraction's design. After entering the Disneyland "space port" and viewing a fifteen-minute briefing film on rocket flight, 102 "space travelers" entered a theater with seats arranged in circular tiers. The ship's progress toward the moon could be tracked through "space scanners" in the floor and on the ceiling, while "sound effects and vibrations gave the 'feel' of an actual aerial journey." First the southern California coastline passed by, and then, after the ship

Kenneth Heuer imagined a "black and mystical forest" on the far side of the moon, with a vile race of beings lurking in the bushes.

Camp on the moon

breached the sound barrier, the noise abated. TWA's Captain Collins informed the passengers that they were now 119,000 miles above the Earth. "To prevent the rocket from hurtling indefinitely into space, the captain must brake the tremendous speed, accomplishing this maneuver by an end-over-end flip so that the rockets are in a position for use as brakes." A sister spaceship, already returning to Earth, could be observed performing the exact same maneuver. The highlight of the trip was a close-up view of the lunar surface, where the captain pointed out the major mountains and valleys. To illuminate the "dark" side, some flares were dropped. Following another end-over-end flip, the "rocket" returned to Earth—"a half-million mile journey accomplished in a quarter of an hour, a preview of the world of the future in Tomorrowland." Walt Disney also produced two related television programs: *Man in Space* and *Man and the Moon.* Both aired in 1955, and Disney himself introduced the programs.

In the 1950s the idea of permanent human settlements on the moon was still an implausible fantasy, but writers devoted considerable imaginative effort to creating such fictive colonies based on the latest scientific developments. In *The Exploration of the Moon* (1954) Arthur C. Clarke made the establishment of a permanent and eventually self-supporting lunar settlement contingent upon the development of nuclear power sources. Just a year before the first actual moon landing took place, Clarke, eccentric as usual, speculated in *The Promise of Space* that the entire moon could become habitable. Because "the most daring prophecies turn out to be laughably conservative," he was convinced that "what the human race will do with the moon during the centuries to come may be as far beyond imagination as the future of the American continent would have been to Columbus." He saw in the moon "a virtual Rosetta Stone that, if properly read, may permit us to learn how the solar system, the Earth and the continents on which we live were formed." Clarke also conjectured that by the year 2020 a trip to the moon would be a distinct possibility for anyone who so desired— "perhaps to see grandchildren who, having been born under lunar gravity, can never come to Earth and have no particular desire to do so. To them it may seem a noisy, crowded, dangerous, and, above all, dirty place." He also imagined "committees of earnest citizens" 150 years later "fighting tooth and nail to save the last unspoiled vestiges of the lunar wilderness." He may yet be proven right on this last point.

Plans to build structures on the moon were by no means limited to the United States. In the second half of the 1950s, besides pursuing their project of interplanetary flights, the Russians declared their intent to build lunar cities with artificial atmospheres to make human life possible. The first so-called Red City was to be erected in a lunar crater under a dome of glass. Sophisticated aluminum doors would form an airlock, and the interior of the city would be further compartmentalized by glass walls with double doors to minimize the possible damage resulting from falling meteorites. These Soviet lunar settlements were expected to be self-sufficient, including locally grown vegetation—which, due to the difference in gravity, could be expected to look dramatically different: "A radish will be as tall as a date palm grows on Earth," declared Nikolai A. Varvarov, the head of the astronautics section of the Russian civil defense organization. "Onions

will send forth sprouts 33 feet long." Since Mars and Venus were envisaged as the next targets, and the moon considered a mere way station, the settlements were pictured as bases for the manufacture of spaceships and the production of fuel. Anatoly Blagonravov from the Soviet Academy of Sciences maintained that "the hoisting of any national flag on the Moon should not be the basic goal." The projected time frame was fifty years, when the moon would have effectively become the seventh continent of planet Earth.

The science journalist Fred Warshofsky chronicled some bizarre speculation about lunar possibilities in *The 21st Century: The New Age of Exploration* (1969). One was the cultivation of algae on human fecal waste under intense sunlight; the algae would be fed to chickens, which would then be consumed by the humans. The moon's reduced gravity would allow fruits and vegetables to grow faster and larger than on Earth. But in a near-perfect vacuum, how could

An ironic illustration of a Kremlin-like structure built under a glass dome on the moon accompanied a 1958 article in *Science Digest*.

a structure filled with breathable air be insulated against leaks? Engineers at the General Electric Company envisioned gouging out a chamber under the surface with nuclear explosives. After decontamination, the cavity would be lined with a strong plastic membrane, then filled with air. This plan was based on blast experiments with pumice rock, which was assumed to resemble the rock of the lunar surface. The astrophysicist Fritz Zwicky (1898–1974) of the California Institute of Technology suggested a way of creating an atmosphere on the moon's surface. First, the moon's mass would have to be increased, either by supplemental waste materials from Earth or by nuclear explosions that would fuse parts of the moon, flatten the lunar mountains, and cut its diameter in half with no lost matter. The gravity would subsequently increase with the smaller surface area so that carbon dioxide could be held in the atmosphere. At another point, Zwicky counseled "to shoot the moon" and to nudge the sun to then observe the consequences from Earth—this was all part of what he called "experimental astronomy," a revolutionary response to humans' history as passive observers.

But secret military scenarios were also being hatched. At the height of the Cold War, in 1958, the U.S. Air Force explored a plan to trigger a nuclear explosion on the moon at least as strong as the one used at Hiroshima. The point of the proposal, made public only decades later, was simple: to display military strength with an atomic mushroom large enough to be visible from the Earth. Leonard Reiffel, the physicist in charge, explained that the bomb was to explode on the lunar edge, so that the ascending cloud would be illuminated by the sun. Reiffel hired the American scientist and astronomer Carl Sagan to calculate how the cloud would expand in space around the moon. The possibility that the explosion might interfere with future scientific research on the moon didn't concern the air force; the public relations triumph justified the means. Although some details of this project ("A119" or "A Study of Lunar Research Flights") remain obscure, with some documents still classified, Reiffel confirmed in 2000 that it was "certainly technically feasible" and could have been accomplished with an intercontinental ballistic missile. Upon the disclosure of the secret plan, David Lowry, a British consultant and nuclear historian, commented, "It is obscene. To think that the first contact human beings would have had with another world would have been to explode a nuclear bomb. Had they gone ahead, we would probably never have had the romantic image of Neil Armstrong taking 'one giant leap for mankind.'"

Scenarios such as this one remind us how thin the line between fact and fiction has always been in the realm of space—thin enough that science fiction could sometimes turn into science fact. But fictional life on the moon often drifted into the realm of the grotesque. The American novelist Robert A. Heinlein wrote *The Moon Is a Harsh Mistress* (1966) when the Apollo program was in full swing, but his moon is not a peaceful one. Instead, it is a forced-labor penal colony in revolt against Earth in a fight for independence. Lunar society of 2075 is characterized in ways that would surprise the most experienced cultural anthropologist: by a form of group marriage and extended families; murder is not prosecuted; insults to women are punishable by death. The dialect of the "Loonies" is a sort of abbreviated English with a strong influence of Russian grammar. Earth rapaciously extracts resources from the moon, where famine and collapse of the colony are imminent. The rebellious Loonies ultimately force Earth to provide recompense for their grain production.

Any attempt to chart the moon in recent imagination has to include Stanley Kubrick's *2001: A Space Odyssey* (1968), which was based on a screenplay by Arthur C. Clarke and premiered in movie theaters just fifteen months before the first moon landing. The film is a speculative reflection on extraterrestrial life and the revolutionary influence of space technology on the future of mankind. It incorporates groundbreaking special effects that continue to amaze viewers. Some critics interpreted it as a critique of modern technology, while others read it as a vehicle for providing a spiritual dimension to the space euphoria of the time. The story revolves around the discovery of a monolith buried below the moon's surface, which, it is assumed, was created by an unknown civilization at some distant point in history.

The late twentieth century brought no return to the lunar utopias of a more optimistic age. In the novel *In Peace on Earth* (1987), by the Polish author and science skeptic Stanisław Lem (1921–2006), the moon fulfills the role of a strange repository. All weapons are transported to the moon to prevent human self-destruction on Earth. But these weapons have the ability to multiply by using the energy of the sun and exploiting the lunar soil.

A lunar mining station is the setting of *Moon* (2008), by the British filmmaker Duncan Jones. The protagonist is an astronaut near the end of a three-year contract supervising the extraction of helium-3, which has become the primary source of energy on Earth. His mission is purely commercial, with no scientific considerations—the astronaut's task is to launch rocket-propelled canisters filled with helium-3 to Earth. The film gives a sense of the alienation experienced in a hostile environment, and the eerie dislocation that might accompany long periods of isolation on the moon. The film's echoes of *2001* and its 1960s-influenced concept of the future give it a retro feel, but its grimy surfaces suggest an even darker flip side to Kubrick's vision.

CHAPTER THIRTEEN

Before and After Apollo

History often rewards great breakthroughs but ignores the preparatory steps that made those achievements possible. The Apollo program, for instance, has been documented in great detail and still receives ample attention, but what of the extraordinary labors that led to that summit? How was flight to the moon realized in practical terms after Jules Verne & Co. designed the blueprints?

When President John F. Kennedy approved the program soon after taking office, he proclaimed that humans would be on the moon before the close of the 1960s. As a result the technicians of the NASA saw their budget suddenly increased tenfold and were able to draw on a tremendous body of technological knowledge. The American public had already embraced the supposedly infinite possibilities of the Space Age. Of course, technological progress doesn't just happen. In an earlier era, the railroad had promised to diminish travel time like no other means of transport had done before. It became a metaphor for the changing relationship of humans to the material world—for the domination of the landscape by technology. In fact, at a time when fast movement was uncommon, travelers found rail travel to be a mind-altering experience, and often a con-

Chandra, the Indian goddess of the moon

229

fusing one. The railroad became an essential asset in the American Civil War, and its influence was still critical to rapid delivery of weapons and matériel during the First World War. Airplanes were quickly put to military use, too, and soon proved highly effective as weapons. In fact, military concerns drove much technological progress in the nineteenth and twentieth centuries.

In the human imagination the way to reach the moon is simply to fly there, but from a purely technical perspective, moon "flight" converged with the history of rocketry at the end of the nineteenth century, when it became clear that rocket propulsion would be needed to transport humans into space. Technology then evolved more or less independently in Russia, the United States, and Germany. Given the backwardness of czarist Russia—an atmosphere that hindered experimentation—it is all the more surprising that Konstantin Eduardovich Tsiolkovsky (1857–1935), a teacher of mathematics, developed what is now seen as the basis of theoretical astronautics. The American Robert Goddard (1882–1945), the Frenchman Robert Esnault-Pelterie (1881–1951), and the German Romanian Hermann Oberth (1894–1989) also independently worked on effective rocket systems.

Tomorrow's geniuses are often today's lunatics, and initially, such endeavors were regarded as esoteric. When Oberth wrote a dissertation on rockets in space and submitted it to the University of Heidelberg, the professors initially rejected it. But by 1923 his thesis, "By Rocket into Planetary Space," inspired wide interest and encouraged further research on the topic of space flight. Oberth was vindicated.

The space program was a pastiche of new and established technologies. As Walter A. McDougall has asserted, it combined "four great inventions: Britain's radar, Germany's ballistic rocket, and the United States' electronic computer and atomic bomb," each "the product of humankind's most destructive conflict—World War II." Immediately after the war, before the start of the space race, these discoveries were applied to the development of intercontinental missiles. The German V-2 rocket program had already established that flight outside the atmosphere was possible. Given the American advance in bombers, the Soviet Union felt particularly compelled to compensate for its lag in warhead technology with a rocket force.

Research on the moon and the idea of flying there attracted people who combined visionary and technical inclinations. Wernher von Braun was a key figure in this effort. From his background as the chief rocket engineer of the Third Reich and a member of the SS, he ultimately became head of launch-vehicle development for the Apollo moon program—an odyssey requiring not only exceptional skills and intelligence but also an opportunistic grasp of diplomacy. As his biographer Michael J. Neufeld has put it, few individuals have "shaken the hand of Eisenhower, Kennedy, Johnson and Nixon—but also Hitler, Himmler, Göring and Goebbels." It

Wernher von Braun, right, on the shooting range of the German Rocket Society in 1930

is known that the German underground rocket factory that produced the V-2 rockets—called Vengeance Weapon 2 by the Nazi Propaganda Ministry—used slave laborers from a nearby concentration camp. It is also known that von Braun was involved in the facility's planning and operation. He clearly benefited from war crimes and bears at least moral if not legal guilt. Even under the most adverse circumstances at the end of a lost war, von Braun managed to control the outcome of events. Realizing that rocket research could best be continued in the United States rather than in Great Britain or Russia, he allowed himself to be captured by the Americans and brought more than one hundred key members of his team along. He was the world's most experienced rocket engineer and soon developed a broader vision of man's venture into space. He suffered a setback in 1954, while working on the advanced rockets program in Huntsville, Alabama, when he was denied the necessary financial support to launch a satellite to ensure that the United States would be the first country in space, but the success of the Soviet Union's Sputnik program boosted his promotion of space exploration.

Humans would not have set foot on the moon so early without the fierce antagonism and competition between the two superpowers during the Cold War. Without the political will and, as Roger D. Launius has put it, "the desire to demonstrate the technological superiority of one form of government over another," the Apollo program would not have been possible. After the USSR launched the *Sputnik 1* satellite in October 1957 and, a month later, *Sputnik 2* with its dog passenger, Laika, the space race gained momentum. Suddenly, Americans found themselves in second place. Because the Soviet Union was known to be working to develop nuclear weapons, its small orbiting orb was seen as a potential military threat to targets inside the United States. The extreme secrecy surrounding the Soviet space program—neither the launch site (the Baikonur Cosmodrome, in the desert of what is now the Republic of Kazakhstan) nor the name of the chief designer of the rockets (Sergei Korolev) was known at that time—certainly contributed to the sense of urgency of the American efforts. This perception translated to a massive infusion of public money into the space race. The Apollo program

cost approximately $25 billion and was driven by Cold War politics rather than important scientific goals. It became, as Launius reminds us, "the largest nonmilitary technological endeavor ever undertaken by the United States."

It is easy to forget that President Kennedy not only failed to define a clear purpose for the program, but also, recognizing the vast costs of the effort, repeatedly tried to persuade Soviet Premier Nikita Krushchev to pursue a joint expedition to the moon—to no avail. The Apollo program had critics from the beginning. According to polls taken in the year leading up the Apollo 11 launch, there was never a clear majority in favor of it. Some objections were based on the belief that "God never intended us to go into space," in the words of one response offered. This is a minority opinion in the population as a whole, but it reminds us of the metaphysical feelings the moon has always evoked. Should this eternal symbol in the sky be touched, its secrets unveiled? In fact, the Apollo program assumed various quasi-religious reverberations. Norman Mailer, for example, characterized the space capsule itself as a sacred object. He also speculated about the extent to which Nazi ideology might have been enmeshed in the

A reunion of pioneers. Wernher von Braun is second from right and Hermann Oberth in the foreground.

space program, mediated through Wernher von Braun and other former German officers.

Most objections, though, had nothing to do with metaphysics and everything to do with economics. As early as 1964, the sociologist Amitai Etzioni characterized the race to the moon as a "monumental misdecision" in his book *The Moondoggle.* The space program, Etzioni argued, produced neither major economic development nor a better understanding of the universe. "Some of the claims are safely projected into a remote and dateless future, others should have never been made; still others exaggerated out of proportion to their real value." All of scientific manpower devoted to space, he wrote, should instead be put into health care or education. "Above all the space race is used as an escape. By focusing on the Moon, we delay facing ourselves, as Americans and as citizens of the Earth." Etzioni later served as senior adviser to the White House under Jimmy Carter. For the eminent science historian and independent humanist thinker Lewis Mumford (1895–1990), the Apollo program was simply a waste of money, "an extravagant feat of technological exhibitionism." He likened the manned space capsule "to the innermost chambers of the great pyramids, where the mummified body of the Pharaoh, surrounded by the miniaturized equipment necessary for magical travel to Heaven, was placed." Still, the Apollo program

Так может выглядеть межпланетный скафандр.

Russian cosmonaut, 1950s

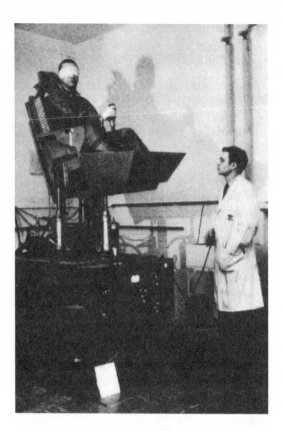

received consistently favorable media coverage. In the wake of the 1968 assassinations of Martin Luther King, Jr., and Robert F. Kennedy, as well as ever more distressing reports from Vietnam, the push toward the moon acted as a counterbalance for the national psyche.

The technological wonders of yesterday rapidly become the banalities of today. The late 1950s and the 1960s held the promise of a world in which nuclear power would be the key to solving problems, replacing scarce energy resources, sharply reducing pollution, and even relieving poverty. The mere threat of nuclear annihilation would put an end to war. Hypersonic air travel would shrink our world while settlements on the moon would expand its capacity. Half a century later, the dreams that became reality have shown a nightmarish side. Implementation of nuclear power has been hamstrung by grave doubts about safety. The supersonic Concorde, a plane developed in 1962 with more than twice the speed of conventional aircraft, was eventually abandoned after a spectacular crash in 2000.

The live transmission of the shaky images of the astronauts' ghostlike steps and their metallic voices from the moon, which triggered a global wave of excitement at the time, has become part of our cultural memory, a relic of a century past. Perhaps the most significant lasting image was that captured when the cameras were turned toward home. For the first time, the entirety of our planet was no longer an abstraction. At last we perceived the Earth in context, not merely as our own specific location. Traveling to the moon gave us an unprecedented sense of our own uniqueness in space and of the limits of the oasis we inhabit. Even as we continued to reach farther outward beyond our world, much of our imagination was turned inward.

Well into the nineteenth century, the moon, understood as a space of the imagination, had assumed a role similar—metaphorically, at least—to those of earlier unknown landscapes and continents. But conquest of the moon was driven by completely different motives than those of the Portuguese and Spanish seafarers in the fifteenth century.

Was the lunar mission really more than a historical accident? It cannot be simply dismissed as a propaganda coup, as an example of hubris, an American ego trip; those characterizations are not false, but they're no more than half-true. The program was not motivated by a desire for wealth, and innovations it inspired still benefit people who never thrilled to Neil Armstrong's "small step." The fuel cell, based on a controlled hydrogen-oxygen reaction, has various applications and may yet prove to be the most useful "clean" alternative to the internal-combustion engine. It was developed in part to power the life support system in a space capsule. Tiny diodes measuring the astronauts' pulse and blood pressure were forerunners of the medical telemetry equipment used today. Freeze-drying made it possible to preserve and to condense such foods as potatoes, peas, carrots, and minced meat, which could later be restored to their original forms with the help of water and a microwave. In addition to these specific developments, Apollo sparked a general interest in engineering, the results of which later benefited various industries. Some of the gadgets and activities associated with moon flight also found their way into popular culture. Michael Jackson's moonwalk was inspired by the astronauts' mode of moving on the lunar surface, and the Italian designer Giancarlo

Zanatto's moonboots would not have made sense without their real-life predecessors.

But to find the most ubiquitous consequence of Apollo-related research, most of us need look no farther than our desktops. The NASA program drove the miniaturization of information technology crucial to the development of modern computers. The Apollo Guidance Computer developed at MIT—a seventy-pound on-board device with a capacity less than that of a cell phone today—made the safe landing on the moon possible. As David A. Mindell writes in *Digital Apollo,* "Apollo began in a world when hardware and electronics were suspect and might fail anytime. It ended with the realization that as electronics became integrated, computers could become reliable, but that software held promise and peril. It could automate the flying, eliminate black boxes from the cramped cabin, and make the subtlest maneuvers seem simple. Yet it was hideously complex and difficult to manage. If it went wrong at a bad time, it could abort a mission or kill its users."

After 1968, about ninety thousand people registered for Pan American World Airways' First Moon Flights Club. Departure was scheduled for the year 2000; Ronald Reagan was one of the first to reserve a seat. The fare for the trip was projected to be fourteen thousand dollars. In retrospect, it is doubtful that a moon flight could have been provided at such a low price, but Pan Am went out of business in 1991 and never had to

Know All Ye by These Presents that

Mr. _____

has become a certified member of Pan Am's

"FIRST MOON FLIGHTS" CLUB

James Montgomery

Number: Vice President, Sales

Pan Am's lunar ambitions

fulfill its lunar obligation. A few years later, en-thusiasm for space exploration had faded, super-seded by more urgent cultural and political issues. America's manned lunar landing project ended in December 1972, when Apollo 17 ended its flight in the Pacific Ocean. In Washington, D.C., NASA officials, astronauts, scientists, and business man-agers celebrated with what the *Washington Post* called "the last splashdown party."

Depending on political perspectives, the story of the U.S. space program can be portrayed as a feel-good triumph or as a waste of money that should have gone to social programs. But there is a third, less common perspective that has stirred considerable emotion over the decades. According to this revisionist narrative, Apollo moon land-ings never happened at all, and what we saw was faked at a terrestrial staging area. According to Roger D. Launius, such theories began to be spun early on, "almost from the point of the first space-flight missions." The explanation offered most commonly for the contrarian view is simple ignorance: for some people this technological accomplishment just didn't fit their worldview, and it was easier for them to imagine a complex hoax than to accept the challenge to their assumptions. For some moon-landing deniers, though, the issue was more complex. They had an almost messianic belief in a conspiracy and often would aggressively resist any discussion of it.

In 1974, at the height of the Watergate scandal, when trust in the American political system was at a low point, Bill Kaysing published *We Never Went to the Moon: America's Thirty Billion Dollar Swindle.* Since then, discussion has raged—often repeating the same arguments ad nauseam. At one point, the former Apollo astronaut Buzz Aldrin was so provoked by Bart Sibrel—who made two films claiming the landings were a fraud and labeled Aldrin "a coward, a liar, and a thief"—that he punched the much younger Sibrel in the face.

Conspiracy theorists disagree about the extent to which the moon landings were faked. A minority believes that the crew actually reached the moon, but that the images were faked to obfuscate the technical details of their journey. Others think that Stanley Kubrick, who directed *2001: A Space Odyssey,* was commissioned by NASA to produce some of the Apollo 11 and 12 footage. According to this scenario, a dummy was launched and allowed to splash down into the ocean. All the other images transmitted into hundreds of millions of homes are said to have been fake footage, produced on a quickly improvised movie set in some remote spot in the Nevada desert. That Kubrick hired former NASA employees for *2001* is, to conspiracy theorists, further evidence. The issue created quite a stir when the Fox television network aired *Conspiracy Theory: Did We Land on the Moon?* in 2001. This hourlong documentary gave a platform to several hoax advocates but offered very little refuting evidence.

Even though a number of independent sources have stressed that the sheer, overwhelming body of physical evidence proves that humans did walk on the moon, the minority view provides admittedly fascinating fodder for publishers, and the nicely packaged story is always guaranteed an audience. If we begin with an open mind, the arguments of the conspiracy theorists may initially seem plausible, but they don't hold up to rigorous scrutiny.

The arguments have been discussed extensively elsewhere, but a taste of the contrarian stance is still in order. Believers in a NASA conspiracy—the Apollo Simulation Project, as some call it—often cite the absence of stars in the jet-black sky photographed by the astronauts. They disregard the explanation that the exposure times were too short to have captured the faint stars. The unusual play of light and shadows in some of the photos, such as the "man on the moon" image of Buzz Aldrin, also draws the ire of conspiracy theorists. Why does it look like a spotlight is directed at him although the sun is clearly behind him or shining from the side? What the conspiracy theorists fail to consider is that the lunar surface has a tendency to reflect light back in the direction of the source. This phenomenon results in a specific glow: a halo or aureole. Often, the deniers are also skeptical about the astronauts' ability to survive exposure to radiation during the trip or the high temperatures on moon's surface. Although radiation and temperature were indeed threats to the astronauts, program scientists charged with astronaut safety clearly resolved the issues. The intensity of denial in the face of all evidence to the contrary reminds us that, for a number of people, the moon landings may have been traumatic. The intrusion of humans onto this numinous orb somehow violated their idea of the natural order. A trip to the moon had been a dream for centuries. Few events are more disturbing than a dream come true.

The issue of the denial also touches upon other, deeper questions. We live in a culture in which the deliberate blurring of boundaries between fact and fiction has achieved a certain artistic legitimacy in popular culture. And while many of us still remember the live images and sounds from the first moon landing, more than half of the world's population is too young to have such an intimate connection with this event. Though not remembering doesn't necessarily translate to denying, they may be less inclined to take its truth for granted.

For those who accept that men have been to the moon, the visits have raised a host of more mundane issues. For example, some people claim to own parts of the moon. But which property laws apply there? And who is authorized to define such laws in the first place? Who—if anyone—should have the right to change the moon's surface? In 1967 the United States, the United Kingdom, and the Soviet Union signed the Outer Space Treaty that declared the moon to be *terra nullius,* a world belonging to no one. About one hundred nations adhere to this agreement today. In 1979 this treaty was supplemented by a more comprehensive one. The Agreement Governing the Activities of States on the Moon and Other Celestial Bodies specifies that the moon should be used for the benefit of all states and peoples and not, for example, as a testing ground for military purposes. The "Moon Treaty" also precludes any state from claiming sovereignty over any territory of a celestial body. However, the treaty has never been ratified by any of the major space-faring powers and remains unsigned by most of them, so it carries no legal weight. Nor is it clear how environmental ethics for the moon can be established and put into practice.

Future projects involving the moon still face major challenges and require piecemeal technological solutions. Companies under contract to NASA are exploring lunar logistics, mining, and spacesuit design. An inflatable test habitat in Antarctica serves as a base for lunar research and a testing station for Mars exploration. Locations such as the Canadian Arctic, the Arizona desert, and the underwater Aquarius habitat serve as key proving grounds. Of course, these extreme terrestrial environments provide only distant approximations of the many challenges that we would confront on future trips to the moon.

The perils of the moon go well beyond the mere terrain. Even though the moon doesn't have the aggressive gases of Venus or radiation contamination of Jupiter, exposure to cosmic rays and solar wind and flares poses much more risk than on Earth, where the atmosphere and the magnetic field act as protective shields. The fine lunar dust, moreover, is highly abrasive and potentially harmful not only to the lungs of the astronauts and to their spacesuits but also to the joints and bearings of lunar robots—a problem that some propose solving by outfitting the robots with disposable coveralls. In any case, robotics will play a major role in lunar exploration, with lightweight vehicles or robotractors moving about the surface and performing tasks that include sampling the lunar soil and extracting any water or oxygen-rich minerals it might contain. Once the water is broken down into hydrogen and oxygen, it might be used to fuel rocket propellants and to create air for breathing.

Champions of the moon's commercial exploitation, such as the Arizona-based Lunar Research Institute, like to stress that even the bulk of the resources needed to build an industrial complex on the moon would not have to be taken there but could be mined on-site. Larry Clark, senior manager for Lockheed Martin's spacecraft technology development laboratory, calculates that processing just the top two inches of soil from an area half the size of a basketball court

A recent
visualization
for a lunar base,
realized by
architects from
the Technical
University of
Darmstadt,
Germany

could yield enough oxygen to keep four astronauts alive for seventy-five days. Silicon can be used to make cells to harvest solar energy, iron to build structures, aluminum, titanium, and magnesium for the construction of spacecraft, and carbon and nitrogen to help grow food. Given the moon's comparably low gravitational pull, transport back to the Earth would be much cheaper than the other way around.

The moon may be a place of infinite possibilities, and some of these projects are reminiscent of the enthusiasm of a gold rush. At the core of these dreams is helium-3, the light variant of a noble gas that is hardly present on Earth but appears to exist in large quantities on the moon. Although Earth's atmosphere and magnetic field shield it from helium-3 in space, the moon lacks these obstacles. Shipped to Earth, the moon's helium-3 could be processed with deuterium in fusion reactors to create helium-4, the energy source of the sun and stars, which could provide power without producing nuclear waste. But such fusion reactors are far from ready for operation; some experts claim that this technology is decades away from commercial viability.

One undeniable resource of the moon is an abundant supply of solar energy, which could be harnessed to power lunar bases or even as an energy source for Earth, thus freeing future generations from dependence on fossil and nuclear fuels. To make the moon's solar energy available on Earth, scientists would have to find a way to convert it into electricity and then into microwaves that could be beamed across space. It is impossible to gauge the effects of sending such concentrations of microwaves through the Earth's atmosphere. Paul D. Spudis, chief of the *Clementine* team (named for a spacecraft launched in 1994 to study the moon's surface by means of special cameras), considers the "Mountain of Eternal Light" near the moon's south pole to be "the most valuable piece of extraterrestrial real estate in the solar system." "The big advantage of this place on the Moon is that it allows you to survive the 14-day lunar night with solar power. You cannot do this on the equator. Moreover, it is close to deposits of excess hydrogen, allowing us to make water, air and rocket propellant. If the Moon is a desert, the poles are its oases," Spudis writes.

All such futuristic projects resonate uneasily

on an Earth that more than ever seems on the brink of ecological disaster. Working on an off-shore oil platform entails many privations, but who would want to spend a substantial part of his or her life on the moon? Who would choose the moon if he or she could have green landscapes, the ocean's shore, and the mountains? Are we really supposed to live somewhere other than Earth? In addition, living in an environment with less gravity than the Earth has severe side effects. A period of rehabilitation is now standard practice after long stays in space. How can detrimental consequences for the circulatory and musculoskeletal systems be limited and psychological and social well-being be maintained? At this point, there still are a lot of open questions.

With no spaceship in sight, the European Space Agency (ESA) can only dream about a manned space mission. Recognizing the immense cost involved, U.S. President Barack Obama has also put the brakes on moon euphoria. The Russians are not in a hurry, either—no cosmonaut will set foot on the moon before 2025. And China doesn't foresee trying to get there before 2030. But there are parts of the world where the notion of a lunar mission still provokes strong emotions.

A case in point is India. When an unmanned spacecraft called *Chandrayaan-1* (moon craft) went to the moon in October 2008, it was a matter of national pride and received extensive coverage in the Indian media. The main aims of this mission were to prepare a three-dimensional atlas in extremely high resolution of both the near and far sides, and to map the lunar surface for the distribution of various elements. For the first time, imaging radar was flown to the moon, resulting in valuable data on the lunar poles. Currently India is planning a manned mission for 2020.

Just before the launch of *Chandrayaan,* millions of Hindu women fasted until the moment when they could first discerned the moon's reflection in a bowl of oil—a rite that they usually perform to safeguard the welfare of their husbands. For many Indians no real contradiction seems to exist between their ancient beliefs and the contemporary scientific pursuits in space. The Indian Space Research Organisation carried the following verse from the *Rig Veda:*

> O Moon!
> We should be able to know you through
>   our intellect,
> You enlighten us through the right path.

Epilogue

Moon Melancholia

From humankind's beginnings, the moon was the mysterious light in the night sky, ascribed sundry magical powers and used to structure time. The telescope changed our view, literally and figuratively. Science asserted the moon's status as a satellite of the Earth, and imagination drifted to speculation about whether anyone—or anything—might live there. Following the Age of Discovery—when indigenous people of the New World were displaced or killed, game was slaughtered, and forests were razed—there was hardly anywhere left on Earth where a utopian society could plausibly have been settled. The moon offered a relatively likely venue for utopia. But it turned out to be lifeless—a blank slate for our dreams, and for our depravity.

In essence, humankind's premodern lunar fantasies were part of the nascent desire to explore outer space, the first toddling steps of the space race. But one of the many consequences of the Copernican revolution was that the moon lost its preeminence in the topography of the sky. No longer considered a planet, it became more remote than ever from its ancient position of honor as one of the seven bodies circling the im-

Lunar contemplation

mobile Earth. Displaced, disenchanted, debased, the moon lost its status. Poets probably sensed first the tremendous change this implied and began to imagine the moon as a shadow world intimately related to Earth, but following another, ultimately unintelligible logic. The speaker in *The Nighttime Chant of a Wandering Asian Sheep-Herder* (1830) by Giacomo Leopardi directs questions at the moon: "What do you do Moon in the sky? Tell me, what do you do,/silent Moon?" His questions remain unanswered. This silence suggests that although the moon is still regarded as a companion and an object of desire, its significance is no longer self-explanatory. The moon has lost its "voice." But even though the moon is unable to answer the restless shepherd's pressing questions, a sense of intimate connection remains, reminiscent of the time when the moon was an important symbol of our connection with the cosmos.

Our most clearly visible neighbor in space, the moon eventually made plausible the possibilities of space travel. With scientific inquiry taking ever-new directions, new objectives emerge. The more complicated life on Earth becomes, the greater the temptation to long for a place where earthly problems do not exist. But perhaps the moon, as the target of some of our hopeful projections, is not all it's cracked up to be. Our future quests may take us to the most unexpected of places. Cultural, spiritual, or scientific progress or renewal may not even be defined in spatial terms, as a movement from one place to another.

In some ways, the interior of our own planet is more remote than the moon, and there is no inherent reason why the future of mankind should be in outer space or why the popular imagination should target some far-away destination in space as its preferred goal. Many scientists are working on uncovering Earth's deep secrets, and the ocean depths still hold many mysteries as well. Many of science's most pressing challenges for the advancement of human knowledge—or even our survival—may be found around the corner rather than in space. If so, will our impulse to use the moon as the screen onto which we project our hopes and fears fade as well?

Over time, our view of which historical events really matter changes. The veteran jet propulsion engineer Wernher von Braun once compared the landing on the moon with the distant moment in natural history when the first animals left the water to find their way on land. The former U.S. president Richard Nixon said that the flight of Apollo 11 represented the most significant week in the history of Earth since the creation. What a difference time makes. Only a few years later, these pronouncements already sounded somewhat inflated, and from the perspective of the twenty-first century, they seem almost absurd overstatements. Boys no longer list "astronaut" as their first choice of profession, and it has become something of a cliché to say that a trip to the moon is no longer an "adventure." The late Carl Sagan even dared call the moon "boring."

In a not-too-distant future, humanity may re-

A comparison of
moon and Earth
at diminished
distance

member the twentieth century for the nuclear
bomb, industrialized genocide, the rising awareness of global warming, the Internet, the achievement of powered flight, and, yes, the first steps
into space—but hardly recall the twelve humans
who actually ventured to the moon. In an effort
to anchor Apollo in memory, in early 2010 a collection of 106 objects left by the Apollo 11 mission—including the bottom stage of the lunar
lander, a seismic monitor, the American flag,
and, ironically, even bags of human waste—were
placed on California's registry of historic land-
marks and resources which makes it the first cultural resource listed not located on Earth.

In our age of moral relativism and multicul-
turalism it has become difficult to speak of eternal truths and universals, but surely the moon
remains an inalienable emblem of the human
imagination—even "in the year 2525," to quote
the lyrics of the 1969 song by Zager and Evans.
There are many moons, but Earth's moon remains something special. But to say it still has
meaning for us is not to say that it can have an
unchanging meaning. Its significance and roles

have always varied across cultures and eras— from heavenly god to symbolic guardian or judge, to the scene or stage of spectacular visions and visits, to being "just" an object of scientific investigation.

Humans have never had to travel to the moon to discover it, or to appreciate its fascination. Why long for the moon if we can see it from the Earth? Does walking on the moon guarantee ultimate access to its secrets? Is it not possible that such physical familiarity may make it impossible to understand those secrets? Technology and water may someday make the moon a habitable place for some humans, but it will—for our lifetimes, at least—remain hostile to life. For the vast majority of us, a trip to the moon will be possible only from an armchair. Maybe it's better that way. Just as we were given a new and unique perspective on Earth from space, perhaps we can grasp the moon only by keeping our distance.

Our future moon of the mind may hold many surprises in store. Unless some unforeseen catastrophe alters its surface or changes the spin or position in space, or our view of it becomes obstructed by some impenetrable layer here on Earth, the moon will continue to *look* the same. But the frameworks from which we see it will inevitably be different; its symbolism is apt to change further as well. Consider, for example, that many scientists consider the moon as a part of the terrestrial system, based on the extensive systemic knowledge we possess about Earth and moon. Thus some observers speak not of the

Earth and its moon but rather of a dual planetary system that revolves around the sun, with the center of mass about three thousand miles from Earth's geographical center. If we accept this view, then the moon becomes essentially another continent.

Perhaps the interpretive frameworks of future generations will be religious, perhaps not, but our relationship with the moon will continue to reflect how we understand our own planet and our place in the universe. Today the moon is not an enigma anymore, but maybe the most surprising fact is that—despite all the knowledge we have about it—the moon is still able to amaze us and create a sense of fascination hardly matched by anything else. *Luna Luna in the Sky. Will You Make Me Laugh or Cry?* read the legend on a T-shirt created by the artist Keith Haring (1958–1990). Haring's question articulates the two faces of the moon in our quest to discover what it means to us. Maybe we should try sometimes to un-think our scientific knowledge of the moon. Just as leaving the city and freeing ourselves from pervasive lighting helps us enjoy the privilege of *seeing* the moon again, putting aside our scientific concepts may help us *perceive* it. Perhaps we can learn from José Arcadio Buendia, the memorable character in Gabriel García Márquez's novel *A Hundred Years of Solitude* who locks himself up with his astronomical instruments for several months to observe the sky, fantasizing about imaginary excursions across unknown oceans. The fictional Buendia deserves a special place among more contemporary stargazers. We may not want to pursue his path with the same degree of tenacity, nor become lost as he did, but it's reassuring to remember that looking up at the moon and stars doesn't have to be purely

"The oldest television"

a matter of astrophysics and mathematics — that a real flight of the imagination may still be possible, even in an age saturated by technology. A final suggestion: the Korean video artist Nam June Paik (1932–2006) called a sculpture he completed in 1967 *Moon Is the Oldest Television*." Why not turn off the box and venture into the night's sky? For while the history of the moon reveals a wealth about our past, maybe even more profound lessons about our present and future await us in its face.

Bibliographic Essay

Titles are mentioned on a selective basis, with focus on references in the English language.

Introduction

For further ideas about life on Earth without the moon see Neil F. Comins, *What if the Moon Didn't Exist? Voyages to Earths That Might Have Been* (New York: HarperCollins, 1993). The "Rare-Earth Hypothesis," positing that the development of life on Earth is the result of improbable circumstances, is also interesting in this context; see Peter Ward and Donald Brownlee, *Rare Earth: Why Complex Life Is Uncommon in the Universe* (Berlin: Springer, 2000).

For a collection on issues related to night in the contemporary world see Paul Bogard, ed., *Let There Be Night: Testimony on Behalf of the Dark* (Reno: University of Nevada Press, 2008).

The International Dark-Sky Association tries to raise awareness of "light pollution." See www.darksky.org.

Chapter One. Gazing at the Moon

A very good resource for many issues concerning the observation of the moon is Patrick Moore, *Patrick Moore on the Moon* (London: Cassell, 2001).

For a more detailed discussion of the question of multiple moons circling Earth see Michael E. Bakich, *The Cambridge Planetary Handbook* (Cambridge: Cambridge University Press, 2000), 145 ff.

The reference for the Shona is taken from William M. Clements, ed., *The Greenwood Encyclopedia of World Folklore and Folklife* (Westport, Conn.: Greenwood, 2005).

For a comprehensive treatment of solar eclipses see J. P. McEvoy, *Eclipse: The Science and History of Earth's Most Spectacular Phenomenon* (London: Fourth Estate, 1999).

The description of the solar eclipse in India is taken from www.krysstal.com/eclipses.html. At the time of writing this Web site also displayed many photos from various solar eclipses.

Camille Flammarion's quotations are taken from *Popular Astronomy: A General Description of the Heavens* (London: Chatto and Windus, 1894). Flammarion is also the source of my quotation of François Arago.

For a general discussion of life outside the Earth in historical perspective see Michael J. Crowe, *The Extraterrestrial Life Debate, 1750–1900: The Idea of a Plurality of Worlds from Kant to Lowell* (Cambridge: Cambridge University Press, 1986).

Richard Holmes, *The Age of Wonder: How the Romantic Generation Discovered the Beauty and Terror of Science* (London: Harper, 2008), has a chapter with biographical information on Sir William Herschel, who started out as an organist in Hannover, Germany, and later became an astronomer who built several hundred telescopes.

Valdemar Axel Firsoff, *Strange World of the Moon: An Inquiry into Its Physical Features and the Possibility of Life* (New York: Basic, 1960), is the most recent book I found in my research that still evaluated the possibility of life on the moon. The perspective of Firsoff, an amateur astronomer, was marginal.

Chapter Two. Moon of the Mind

The role of the moon, the sun, and stars in ancient cultures fills volumes. A comprehensive overview of the historical relationship of humans toward the cosmos can be found in:

John North, *Cosmos: An Illustrated History of Astronomy and Cosmology* (Chicago: University of Chicago Press, 2008); Edwin C. Krupp, *Beyond the Blue Horizon: Myths and Legends of the Sun, Moon, Stars, and Planets* (New York: HarperCollins, 1991); and Anthony Aveni, *People and the Sky: Our Ancestors and the Cosmos* (London: Thames and Hudson, 2008). Aveni's older book *Empires of Time: Calendars, Clocks, and Cultures* (New York: Basic, 1989), is also relevant for some of the issues dealt with in this chapter.

Peter Watson's massive *Ideas: A History of Thought and Invention from Fire to Freud* (London: Weidenfeld and Nicolson, 2005) deals with some of the aspects of this chapter in the broader context of a more general history of human ideas and inventions.

The examples for lunar myths of the Maoris, the Tupí, and the Tartars, as well as the various mythical tropes, are taken from Jules Cashford, *The Moon: Myth and Image* (New York: Four Walls Eight Windows, 2003).

On the cult of the moon in Lusitania see Moisés Espírito Santo, *Cinco mil anos de cultura a oeste. Etno-história da religião popular numa região da Estremadura* (Lisbon: Assírio and Alvim, 2004).

For more scholarly information on the significance of the sun, the moon, and the stars in ancient India refer to Georg Feuerstein, Shubhash Kak, and David Frawley, *In Search of the Cradle of Civilization: New Light on Ancient India* (Wheaton: Quest, 2001).

The reference to Roger Bacon's measurement of the distance between Earth and moon is taken from Albert van Helden, *Measuring the Universe: Cosmic Dimensions from Aristarchus to Halley* (Chicago: University of Chicago Press, 1985).

Martin Nilsson, *Primitive Time-Reckoning: A Study in the Origins and First Development of the Art of Counting Time Among the Primitive and Early Culture Peoples* (Lund: Gleerup, 1920), remains a fascinating study of early timekeeping systems.

The information about the Hopi Indians' astronomical

system is taken from Stephen C. McCluskey, "The Astronomy of the Hopi Indians," *Journal for the History of Astronomy* 8 (1977): 174–195.

Chapter Three. Charting the Moonscape

Comprehensive treatments of the evolution of lunar maps are offered by Ewen Adair Whitaker, *Mapping and Naming the Moon: A History of Lunar Cartography and Nomenclature* (Cambridge: Cambridge University Press, 1999) and Scott L. Montgomery, *The Moon and the Western Imagination* (Tucson: University of Arizona Press, 1999). A major part of my treatment in this chapter is based on these books. A more technical resource with a somewhat different selection of historical maps is Zdeněk Kopal and Robert W. Carder, *Mapping the Moon: Past and Present* (Dordrecht: D. Reidel, 1974).

Galileo Galilei, *Sidereus Nuncius; or, The Sidereal Messenger,* trans. Albert van Helden (Chicago: University of Chicago Press, 1989).

The quotation of Martin Kemp related to Galileo is taken from *Seen/Unseen: Art, Science, and Intuition from Leonardo to the Hubble Telescope* (Oxford: Oxford University Press, 2006).

A modern photographic map of the moon, including the locations of the Apollo landings, can be found at www.google.com/moon/.

Chapter Four. Pale Sun of the Night

On the issue of the moon's whiteness: Sobha Sivaprasad and George M. Saleh, "Why Is the Moon White?"; Robert W. Kentridge, "Constancy, Illumination, and the Whiteness of the Moon"; Robert W. Kentridge and Paola Bressan, "The Dark Shade of the Moon"; and David H. Foster, "Confusing the Moon's Whiteness with its Brightness," all in *Clinical and Experimental Ophthalmology* 33 (2005): 571–575.

The quotation of Leonardo da Vinci is from *Philosophical Diary,* trans. Wade Baskin (New York: Philosophical Library, 1959).

The Alexander von Humboldt quotation is from *Cosmos: A Sketch of a Physical Description of the Universe,* trans. E. C. Otte and B. H. Paul (New York: Harper and Brothers, 1868).

On the historical evolution of the measurement of light intensity see J. B. Hearnshaw, *The Measurement of Starlight: Two Centuries of Astronomical Photometry* (Cambridge: Cambridge University Press, 1996).

The quotation about moonless nights in ancient Rome is taken from Jérôme Carcopino, *Daily Life in Ancient Rome: The People and the City at the Height of the Empire,* trans. E. O. Lorimer (New Haven: Yale University Press, 2003).

The source of the anecdote on Jacqueline Kennedy Onassis and the Taj Mahal is Wayne Koestenbaum, *Jackie Under My Skin: Interpreting an Icon* (New York: Farrar, Straus and Giroux, 1995).

The Johann Wolfgang von Goethe quotation is from *The Autobiography of Goethe: Truth and Poetry from My Own Life,* trans. John Oxenford (London: Bell and Daldy, 1867).

The quotation of Henry Matthews can be found in *The Diary of an Invalid; being the Journal of a Tour in Pursuit of Health; in Portugal, Italy, Switzerland, and France in the years 1817, 1818, and 1819* (London: John Murray, 1820).

The quotation of Doreen Valiente is from *Where Witchcraft Lives* (London: Aquarian, 1962).

The quotations of Marcel Proust are from *In Search of Lost Time (Remembrance of Things Past),* trans. C. K. Scott Moncrieff and Terence Kilmartin (New York: Random House, 1981).

For a vivid account of the members of the Lunar Society see Jenny Uglow, *The Lunar Men: Five Friends Whose Curiosity Changed the World* (New York: Farrar, Straus and Giroux, 2002).

A recent study of the moon in Italian culture is Pietro

Greco, *L'astro narrante. La Luna nella scienza e nella lettera-tura italiana* (Milan: Springer-Verlag Italia, 2009).

For a study of Yoshitoshi's moon illustrations see John Stevenson, *Yoshitoshi's One Hundred Aspects of the Moon* (Leiden: Hotei, 2001).

On the specifics of how to create a moon garden, see the popular guide by Marcella Shaffer, *Planning and Plant-ing a Moon Garden* (North Adams, Mass.: Storey, 2000).

For other cultural aspects of the night see the following works:

Christopher Dewdney, *Acquainted with the Night: Excur-sions Through the World After Dark.* (London: Bloomsbury, 2004).

Roger A. Ekirch, *At Day's Close: Night in Times Past* (New York: Norton, 2006).

Eluned Summers-Bremner, *Insomnia: A Cultural History* (London: Reaktion, 2008).

Chapter Five. Encounters of a Lunar Kind

Charles Morton (1627–1698), an English naturalist, in a remarkable if strange combination of astrology and orni-thology, suggested in all earnestness that birds hibernate on the moon. As Thomas B. Harrison has remarked, this was "the earliest treatise on bird migration on England." Morton seems to have been inspired by Godwin's novel—the rather unusual case of an idea taken from literature making its way into science: Thomas B. Harrison. "Birds in the Moon," *Isis* 45 (1954): 323–330.

The following monograph helped me with identifying some of the English and French examples in this chapter: Stephan Edinger, *Literarische Reisen zu fernen Planeten. Eine ideengeschichtliche Untersuchung zur französischen Literatur des 19. Jahrhunderts* (Marburg: Tectum Verlag, 2005).

Here are the original titles and translations in the order they appear in the chapter. Not all of these works have been translated into English.

Bernard De Fontenelle, *Entretiens sur la pluralité des mondes* (Discourses on the plurality of worlds).

Ludovico Ariosto, *Orlando furioso* (The frenzy of Orlando).

Cyrano De Bergerac, *L'Autre Monde: Histoire comique des états et empires de la lune* (The other world: The comical history of the states and empires of the moon).

Madam la Baronne de V***, *Le char volant* (The flying tank).

Anonymous, *Histoire intéressante d'un nouveau voyage à la Lune* (Interesting account of a new trip to the moon).

Alexandre Cathelineau, *Voyage à la Lune* (Trip to the moon).

Jacques Bujault, *Voyage dans la Lune* (Trip to the moon).

Anonymous, *Voyage tout récent dans la lune* (A very recent trip to the moon).

Louis Desnoyer, *Les Aventures de Robert Robert* (The adventures of Robert Robert).

De Sélènes, Pierre, *Un monde inconnu. Deux ans sur la lune* (An unknown world: Two years on the moon).

The classic survey of literary trips to the moon is Marjorie Hope Nicolson, *Voyages to the Moon* (New York: Macmillan, 1948).

A more recent treatment is Aaron Parrett, *The Translunar Narrative in the Western Tradition* (Aldershot and Burlington: Ashgate, 2004).

Lucian Boia's *L'Exploration imaginaire de l'espace* (Paris: Editions La Découverte, 1987) is an amazing collection of literary travels into space and has also been useful in writing this chapter.

The Russian examples are taken from Stephen Lessing Baehr, *The Paradise Myth in Eighteenth-Century Russia: Utopian Patterns in Early Secular Russian Literature and Culture* (Stanford: Stanford University Press, 1991).

Chapter Six. Lunar Passion in Paris

Jules Verne, *From the Earth to the Moon,* 1865 (available in various editions).

William Butcher, *Jules Verne. The Definitive Biography* (New York: Thunder's Mouth, 2006).

Chapter Seven. Accounts of Genesis

For the theories of Darwin and Urey see Stephen G. Brush, *Fruitful Encounters: The Origin of the Solar System and of the Moon from Chamberlin to Apollo* (Cambridge: Cambridge University Press, 1996).

Dana Mackenzie narrates the quest to explain the genesis of the moon in her formidable *The Big Splat, or How Our Moon Came to Be: A Violent Natural History* (Hoboken: Wiley, 2003).

On the possibility of a metallic core of the moon see Ian Garrick-Bethell, Benjamin P. Weiss, David L. Shuster, and Jennifer Buz, "Early Lunar Magnetism," *Science* 323 (2009): 356–359.

Chapter Eight. A Riddled Surface

James Lawrence Powell's readable *Mysteries of Terra Firma: The Age and Evolution of the Earth* (New York: Free Press, 2001) presents a detailed introduction to the question of the origin of and theories about the lunar craters.

For a more comprehensive treatment of the moon from a geological perspective see Stuart Ross Taylor, "The Moon," in *Encylopedia of the Solar System,* 2nd ed., ed. Lucy-Ann McFadden, Paul R. Weissman, and Torrence V. Johnson (San Diego: Elsevier Science/Academic Press, 2007).

For the discussion of Aepinus see Roderick Home, "The Origin of the Lunar Craters: An Eighteenth-Century View," *Journal for the History of Astronomy* 3 (1972): 1–10.

The quotation of Paul D. Spudis is taken from *The Once*

and Future Moon (Washington, D.C.: Smithsonian Institution Press, 1996).

The quotations of Richard Anthony Proctor are from *The Moon: Her Motions, Aspect, Scenery, and Physical Condition* (New York: Appleton, 1886).

On the latest discoveries regarding water on the moon see, e.g., Kenneth Chang, "Water Found on Moon, Researchers Say," *New York Times,* November 14, 2009.

The exotic theories of Peal, Fountain, Fauth, Davis, and Ocampo related to the makeup of the lunar surface are taken from the chapter "How the Lunar Craters *Weren't* Formed" in Patrick Moore, *The Wandering Astronomer* (Bristol: Institute of Physics Publishing, 2003).

Chapter Nine. Lunar Choreography

On the possible influence of the moon on seismic activity see "Can the Moon Cause Earthquakes?" *National Geographic News,* May 23, 2005.

On the research into the condition of trees and wood at different times of the year see Ernst Zürcher et al., "Looking for Differences in Wood Properties as a Function of the Felling Date: Lunar Phase Correlated Variations in the Drying Behavior of Norway Spruce (*Picea abies* Karst.) and Sweet Chestnut (*Castanea sativa* Mill.)," *Trees: Structure and Function,* Springer, August 26, 2009 (online publication).

The references for studies on seals and skylarks are: P. Watts, "Possible Lunar Influence on Hauling-out Behavior by the Pacific Harbor Seal (*Phoca vitulina Richardsi*)," *Marine Mammal Science* 9 (1993): 68–76; D. James, G. Jarry and C. Érard, "Effet de la lune sur la migration postnuptiale nocturne de l'alouette des champs *Alauda arvensis L.* en France," *Sciences de la vie* 323 (2000): 215–224.

On the mass spawning of corals see William J. Broad, "Sexy Corals Keep 'Eye' on Moon, Scientists Say," *New York Times,* October 19, 2007.

On the reproductive behavior of sea urchins and its synchronization with lunar phases see Bill Kennedy and John S. Pearse, "Lunar Synchronization of the Monthly Reproductive Rhythm in the Sea Urchin *Centrostephanus coronatus* Verrill," *Journal of Experimental Marine Biology and Ecology* 17 (1975): 323–331.

On the dung beetle and the role of the moon as visual landmark see Eric Warrant and Dan-Eric Nilsson, eds., *Invertebrate Vision* (Cambridge: Cambridge University Press, 2006).

Chapter Ten. Esoteric Practices

The reference for the account of the moon doctor is M. Herz, "Die Wallfahrt zum Monddoktor in Berlin," *Berlinische Monatsschrift* (1783), 368–385.

The account of medical astrology is mostly based on Mark Harrison, "From Medical Astrology to Medical Astronomy: Sol-lunar and Planetary Theories of Disease in British Medicine, c. 1700–1850," *British Journal for the History of Science* 33 (2000): 25–48.

The quotations of Erasmus Darwin are from *Zoonomia; or, The Laws of Organic Life,* 2 vols. (London: J. Johnson, 1794-1796).

For a study of pagan ritual from a historical perspective see Ronald Hutton, *The Triumph of the Moon: A History of Modern Pagan Witchcraft* (Oxford: Oxford University Press, 1999).

The details of the description of the Wiccan full-moon ritual are taken from http://paganwiccan.about.com/od/moonphasemagic/ht/SpringFullMoon.htm.

Chapter Eleven. Spurious Correspondences

Folklore related to the moon is ubiquitous, and often the tales are surprisingly similar across cultures. The examples of superstitions in German folklore are taken from Werner Wolf, *Der Mond im deutschen Volksglauben* (Bühl: Druck und Verlag der Konkordia AG, 1929). The source of the examples from the Philippines is Francisco R. Demetrio, *Encyclopedia of Philippine Folk Beliefs and Customs* (Cagayan de Oro City: Xavier University, 1991).

For the described case study of the lycanthropic woman see Harvey Rosenstock and Kenneth R. Vincent, "A Case of Lycanthropy," *American Journal of Psychiatry* 134 (1977): 195–205.

On the werewolf issue see Ian Woodward, *The Werewolf Delusion* (New York: Paddington, 1979).

On the topic of the alleged impact of the moon on human behavior also see the following articles and studies mentioned in the chapter:

Scott O. Lilienfeld and Hal Arkowitz. "Lunacy and the Full Moon: Does a Full Moon Really Trigger Strange Behavior?" *Scientific American Mind,* February 9, 2009.

I. W. Kelly, James Rotton, and Roger Culver, "The Moon Was Full and Nothing Happened: A Review of Studies on the Moon and Human Behavior and Human Belief," in *The Outer Edge,* ed. J. Nickell, B. Karr, and T. Genoni (Amherst, N.Y.: CSICOP, 1996).

J. M. Gutiérrez-García and F. Tusell, "Suicides and the Lunar Cycle," *Psychological Reports* 80 (1997): 243–250.

The Swiss study mentioned is Martin Röösli et al., "Sleepless Night, the Moon Is Bright: Longitudinal Study of Lunar Phase and Sleep," *Journal of Sleep Research* 15 (2006): 149–153.

These are the studies concerned with birth and fertility: E. Periti and R. Biagiotti, "Lunar Phases and Incidence of Spontaneous Deliveries: Our Experience," *Minerva Ginecologica* 46 (1994): 429–433; J. M. Arliss, E. N. Kaplan, and S. L. Galvin, "The Effect of the Lunar Cycle on Frequency of Births and Birth Complications," *American Journal of Obstetrics and Gynecology* 192 (2005), 1462–1464.

On the menstrual patterns among the Dogon see B. I. Strassmann, "The Biology of Menstruation in *Homo sapiens:* Total Lifetime Menses, Fecundity, and Nonsynchrony in a Natural Fertility Population," *Current Anthropology* 38

(1997): 123–129; Sung Ping Law, "The Regulation of Menstrual Cycle and Its Relationship to the Moon," *Acta Obstet Gynecol Scan* 65 (1986): 45–48; T. B. Criss and J. P. Marcum, "A Lunar Effect on Fertility," Social Biology 28 (1981): 75–80.

For a current scientific discussion of human biological processes and calendrical cycles see R. G. Foster and T. Roenneberg, "Human Responses to the Geophysical Daily, Annual, and Lunar Cycles," *Current Biology* 18 (2008): 784–794.

On the historical roots of belief in the power of the moon to cause insanity and epilepsy see Charles A. Raison, Haven M. Klein, and Morgan Steckler, "The Moon and Madness Reconsidered" *Journal of Affective Disorders* 53 (1999): 99–106.

The following books explore the field of chronobiology:

Russel G. Foster and Leon Kreitzman, *Rhythms of Life: The Biological Clocks That Control the Daily Lives of Every Living Thing* (New Haven: Yale University Press, 2005).

Russel G. Foster and Leon Kreitzman, *Seasons of Life: The Biological Rhythms That Living Things Need to Thrive and Survive* (New Haven: Yale University Press, 2009).

Klaus-Peter Endres and Wolfgang Schad, *Moon Rhythms in Nature: How Lunar Rhythms Affect Living Organisms,* trans. Christian von Armin (Edinburgh: Floris, 2002). This book relies on the tradition of anthroposophy, which has traditionally been concerned with the study of "rhythmical processes" in nature.

The Bob Berman quotation is from *Secrets of the Night Sky: The Most Amazing Things in the Universe You Can See with the Naked Eye* (New York: William Morrow, 1995).

Chapter Twelve. Visions of the Moon

For a detailed analysis of James Nasmyth's photographs see Frances Robertson, "Science and Fiction: James Nasmyth's Photographic Images of the Moon," *Victorian Studies* 48 (2006): 595–623.

The description of A Trip to the Moon in Coney Island's Luna Park is based on a text by Jeffrey Stanton at www .westland.net/coneyisland/articles/lunapark.htm.

The original German title of Karl-August Laffert's novel is *Der Untergang der Luna* (Berlin: Stilke, 1921).

For a more general discussion of technology in American culture, including the role of world's fairs to promote space travel, see David E. Nye, *American Technological Sublime,* (Cambridge: MIT Press, 1994).

For a thorough discussion of science in movies see Sidney Perkowitz, *Hollywood Science: Movies, Science, and the End of the World* (New York: Columbia University Press, 2007).

The full reference for the text about the vile race of lunarians on the far side of the moon: Kenneth Heuer, *Men of Other Planets* (New York: Pellegrini and Cudahy, 1951). My thanks go to the staff of the Strand Bookstore for pointing this out to me.

The information for the segment about the plan for a nuclear bombing of the moon is from Antony Barnett, "U.S. Planned One Big Nuclear Blast for Mankind," *Observer,* May 14, 2000.

The description of the moon ride in Disneyland is based on "Feeling of Space Journey Accompanies Trip to Moon," *Disneyland News,* January 1957.

At the time of writing, the movies by Georges Méliès, Fritz Lang, and Arthur Hilton, as well as the Walt Disney program *Man and the Moon,* were available on YouTube. com. In one of the sections of the latter program Wernher von Braun explains the principles of rocket propulsion.

For an interesting introduction to space-age euphoria by one of its major protagonists see Arthur C. Clarke, *The Promise of Space* (New York: Harper and Row, 1968). For ideas on lunar stations see also Fred Warshofsky, *The 21st Century: The New Age of Exploration* (New York: Viking, 1969).

On Fritz Zwicky and his "experimental astronomy": Richard Panek, "The Father of Dark Matter Still Gets No

Respect," *Discover,* January 2009, http://discovermagazine
.com/2009/jan/30-the-father-of-dark-matter-still-gets-
no-respect.

Chapter Thirteen. Before and After Apollo

On the effect of railway travel on the human imagination
see Wolfgang Schivelbusch, *Railway Journey: The Industrial-
ization of Time and Space in the 19th Century* (Berkeley: University of California Press, 1986).

The quotation of Walter A. McDougall is from *The
Heavens and the Earth: A Political History of the Space Age*
(New York: Basic, 1985).

The quotation of Michael J. Neufeld about Wernher
von Braun is from *Von Braun: Dreamer of Space, Engineer of
War* (New York: Knopf, 2007).

Space travel, including rocketry development, is the
subject of: Wernher von Braun and Frederick I. Ordway
III, *History of Rocketry and Space Travel* (New York: Crowell,
1975).

Frederick I. Ordway III, *Blueprint for Space: Science Fiction to Fact* (Washington, D.C.: Smithsonian Institution
Press, 1992).

Howard E. McCurdy, *Space and the American Imagination* (Washington, D.C.: Smithsonian Institution Press,
1997).

For more scholarly treatments of space travel and the
Apollo program see Roger D. Launius, *Frontiers of Space
Exploration* (Westport, Conn.: Greenwood, 2004), and
Everett C. Dolman, *Astropolitik: Classical Geopolitics in the
Space Age* (London: Frank Cass, 2001). And, by the same
author, another essay dealing with the various historical
spaceflight narratives and with the issue of conspiracy in
the context of the moon landings: "American Spaceflight
History's Master Narrative and the Meaning of Memory,"
in *Remembering the Space Age: Proceedings of the 50th Anni-
versary Conference,* ed. Steven J. Dick (Washington, D.C.:
NASA, 2008), www.asiaing.com/remembering-the-space-

age-proceedings-of-the-50th-anniversary-conference.html (a free e-book).

For a scholarly survey of Russian space history see Brian Harvey, *Russian Planetary Exploration: History, Development, Legacy, Prospects* (Berlin: Springer, 2007).

Books about the Apollo program and spaceflight abound. Here is a small selection:

Andrew Chaikin, *A Man on the Moon: The Voyages of the Apollo Astronauts* (New York: Viking, 1994).

Bob Berman, *Shooting for the Moon: The Strange History of Human Spaceflight* (Guilford, Conn.: Lyons, 2007).

Craig Nelson, *The Epic Story of the First Men on the Moon* (New York: Viking, 2009).

Gerard J. DeGroot, *Dark Side of the Moon: The Magnificent Madness of the American Lunar Quest* (New York: New York University Press, 2006).

On the integration of man and computer systems in the context of Apollo, and on various other, more technical issues and questions, such as the need for human presence in future space exploration, see David A. Mindell, *Digital Apollo: Human and Machine in Spaceflight* (Boston: MIT Press, 2008).

This book deals in more detail with the idea that the lunar landings never happened: Philip C. Plait, *Bad Astronomy: Misconceptions and Misuses Revealed, from Astrology to the Moon Landing "Hoax"* (New York: Wiley, 2002).

Several sites on the Internet claim that humans never landed on the moon, for example: www.moonmovie.com.

Roger D. Launius has written an interesting article about the ways in which the history Apollo program has been told, depending on political inclinations: "American Spaceflight History's Master Narrative and the Meaning of Memory."

Information about future lunar stations can be found in Alan Boyle, "Wanted: Home-builders for the Moon. NASA's Post-2020 Plan Involves the Usual (and Unusual) Space Suspects," MSNBC, February 1, 2007.

On the physical consequences of space travel see

Jerome Groopman, "Medicine on Mars: How Sick Can You Get During Three Years in Deep Space?" *New Yorker,* February 14, 2000, 36–41.

On the topic of water on the moon see Marc Chaussidon, "Planetary Science: The Early Moon Was Rich in Water," *Nature* 454 (2008): 170–172.

References to Lewis Mumford are from Donald L. Miller, *Lewis Mumford: A Life* (New York: Grove, 2002).

The astronautical engineer Robert Zubrin's *Entering Space: Creating a Spacefaring Civilization* (New York: Jeremy P. Tarcher/Putnam, 1999) is a passionate plea for the colonization of space, financed by private enterprise. It includes ideas for colonizing Mars, for mining the solar system, and for building fusion-powered spaceships. A helium-3 mining facility on the moon is one of the first steps.

On the Indian space mission see Tunku Varadajan, "Fly Me to the Deity," *New York Times,* October 29, 2008.

For a thorough discussion of how the view of our planet from space changed our view of the Earth see Robert Poole, *Earthrise: How Man First Saw the Earth* (New Haven: Yale University Press, 2008).

Epilogue

Richard Nixon made his statement on July 20, 1969.

While I was completing this manuscript, Barack Obama was calling on NASA to end the program that was designed to bring humans back to the moon and focus instead on radically new space technologies.

In addition to the above references, the following books have been helpful with a number of aspects:

Marta Erba, Gianluca Ranzini, and Daniele Venturoli, *Dalla Luna alla Terra. Mitologia e realtà degli influssi lunari* (Turin: Bollati Boringhieri, 2010).

Daniel Grinsted, *Die Reise zum Mond. Zur Faszinationsgeschichte eines medienkulturellen Phänomens zwischen Realität und Fiktion* (Berlin: Logos, 2009).

Acknowledgments

I would like to thank above all my editor Jean E. Thomson Black for her continuing enthusiasm and support, Dan Heaton for his intensive and thoughtful work on the final manuscript, Sonia Shannon for designing this book, Jaya Chatterjee for helping me with the organization of the illustrations, and the staff of Yale University Press in the New Haven and London offices. Particular thanks are due to David Luljak for his index. Many thanks also to my parents, Helgard and Siegfried Brunner, and to the following friends and associates: Rathnayake M. Abeyrathne (University of Peradeniya, Kandy, Sri Lanka), Ines Cabral, Alexandre Cabrita, José João Dias Carvalho, Francesco Paolo de Ceglia (University of Bari Aldo Moro, Italy), Dan Delany, Detlef Feussner, Jacinto José Gomes, Joseph P. Grubb, Michael Hely, Klaas Jarchow, Brendan Kenney, Lori Lantz, Ulrich Meyer, Olaf Oberschmidt, Scott W. Perkins, Eva Schoening, Julien Sialelli, Benjamin A. Smith, Keijiro Suga (Meiji University, Tokyo, Japan), and a very dear friend who prefers to remain unnamed here.

My gratitude also extends to the many often unknown artists whose fine illustrations are featured in this book and—for a range of reasons—to Margaret Adamic (Disney Publishing Worldwide, Inc.), Paul Brown (National Park Service), Jules Cashford, Nina Cummings (The Field Museum, Chicago), Volker Dehs, Ian Garrick-Bethell (MIT), Pietro Greco, Jeannine Green (Bruce Peel Special Collections Library, University of Alberta), Daniel Grinsted, Ove Hoegh-Guldberg (Global Change Institute, University of

Queensland), Eric van den Ing, Edwin C. Krupp (Griffiths Observatory, Los Angeles), Jacques Laskar (Observatoire de Paris), Benjamin Lazier (Reed College, Portland), Vera Martinez (Technical University Darmstadt), Nancy O'Shea (The Field Museum, Chicago), Jeff Papineau (Bruce Peel Special Collections Library, University of Alberta), Bernd A. Pflumm, Marco Scola, Paul D. Spudis (Lunar and Planetary Institute, Houston), Stuart Ross Taylor (Australian National University, Canberra), Ana Tipa, Joe Tucciarone, Mark Wieczorek (Institut de Physique du Globe de Paris), the librarians of the Berlin State Library, the New York Public Library, and the National Library of Portugal in Lisbon, as well as several sellers of antique books, especially Christoph Janik. I would also like to thank the readers who evaluated the manuscript and gave constructive criticism.

A number of books have been influential in my writing; the most important of these are included in the bibliographic essay.

Any problems that remain are, of course, my own responsibility.

If you want to drop me a line, please do so by E-mail: bbrunner@gmx.net.

Illustration Credits

Index